Innovation, Complexity and Policy

Matthias Weber (ed.)

Innovation, Complexity and Policy

Contributions from 30 years of innovation policy research in Austria

PL ACADEMIC RESEARCH

Bibliographic Information published by the Deutsche Nationalbibliothek
The Deutsche Nationalbibliothek lists this publication in the Deutsche Nationalbibliografie; detailed bibliographic data is available in the internet at http://dnb.d-nb.de.

Library of Congress Cataloging-in-Publication Data
Names: Weber, Karl Matthias, 1966- editor.
Title: Innovation, complexity and policy / Matthias Weber (ed.).
Description: Frankfurt am Main ; New York : Peter Lang, [2017]
Identifiers: LCCN 2017017249 | ISBN 9783631723159
Subjects: LCSH: Technology and state—Austria—History. | Technological innovations—Austria—History. | Technological innovations—Government policy—Austria—History.
Classification: LCC T26.A9 I56 2017 | DDC 338.9436/06--dc23 LC record available at https://lccn.loc.gov/2017017249

Cover Image: Coop Himmelblau Building in Seibersdorf
©Gerald Zugmann/ www.zugmann.com

ISBN 978-3-631-72315-9 (Print)
E-ISBN 978-3-631-72316-6 (E-PDF)
E-ISBN 978-3-631-72317-3 (EPUB)
E-ISBN 978-3-631-72318-0 (MOBI)
DOI 10.3726/b11137

PL Academic Research is an Imprint of Peter Lang GmbH.

Peter Lang – Frankfurt am Main · Bern · Bruxelles · New York · Oxford · Warszawa · Wien

This publication has been peer reviewed.

www.peterlang.com

Table of Contents

Ingolf Schädler

Foreword

This volume is dedicated to the valued scientist, colleague and good friend, Sepp Fröhlich, on the occasion of his retirement. It is time to pause and pay tribute to a major figure in Austrian technology and innovation policy.

I know that Sepp would already be contradicting me by now. Being a humble man as he is, he certainly won't like the term tribute. But there are also other reasons: he would not find the phrase Austrian technology and innovation policy – and the word order here is deliberately chosen – sufficiently broad. And it doesn't do him justice either, even though it still remains a cause close to his heart. More important are the efforts Sepp has made to provide its theoretical foundations or, more precisely, its theoretical modelling. In his eyes, every form of technology policy should have knowledge-based, theoretical foundations. Otherwise it is at best only "policy", with exactly this connotation, of course. So deep in his heart, he always felt to be a scientist than anything else.

And so we met, and learned to respect one another, while trying to develop the first theoretical model for an Austrian technology policy in the 1980s – the term "innovation policy" did not yet exist in the public consciousness at that time. And thus the Austrian Technology Monitoring System, ATMOS, was born. And lots of people joined in; from the ministries, TU Wien, the Austrian Institute of Economic Research WIFO, Joanneum Research. Sepp and his team from the former Austrian research centre at Seibersdorf (ÖFZS) were right there in the midst of it all.

We were the young guns of Austrian technology policy. And we failed completely with the ATMOS model. Nevertheless, it gave rise to something new. Something which could be called the start of a systematic Austrian technology policy. A network of young technology policy specialists, closely wedded to science and rationality. They quickly took up positions in several not unimportant institutions where they have had a significant impact on Austrian and international technology policy to the current day.

It was an exciting time. Bruno Kreisky, the Austrian prime minister of that time, has opened doors to the outside world. He inspired a whole generation in universities and research centres to challenge the post-war science system. There was a spirit of optimism in the air, in Austria, in Europe, and also in technology policy. We had all read our Jean-Jacques Servan-Schreiber thoroughly: The

American Challenge, a book which earned international acclaim in the late 1970s and clearly alarmed the European continent, then still divided into the EEC and EFTA. It focused on the economic threat to Europe posed by America's increasingly dominant technological leadership in many knowledge domains at the time. How little has changed!

Many countries, Austria amongst them, responded by drawing up their own technology programmes. Sepp and his team at ÖFZS, which later became the Austrian Institute of Technology, played a central role in the preparation, design and ex-post evaluation of these technology programmes. Today Austria need not shy away from international comparisons, although this was not always the case.

As the result of an initiative by former French President Mitterrand, the EEC and EFTA agreed to jointly cooperate in the technology sector at European level. EUREKA was born, and the cooperation between the European states on the European Union's innovation and technology programmes which started at that time continues to the present day. Today Sepp's team is amongst the most important advisors to the European Commission on matters concerning technology foresight, completing the picture of what had once started out as knowledge-based technology policy.

Much has changed since then. Austria has modernised. It has become more research and technology-intensive, with one of Europe's highest research intensities at over 3 %, a strongly export-focused industry, a positive technology balance, and a high rate of patent registrations. Many of the models we developed back then, such as the Austrian competence centre programmes, are now internationally regarded as models of best practice and copied many times over.

Austria's role in European technology policy is also widely recognised at European level. Austria has clearly benefitted from the modernisation of the past 20 years, triggered by its EU membership and hugely successful participation in the EU's RTI programmes. Austria has long since shed its image as a latecomer.

Much has been achieved, but much still remains to be done. Nevertheless, it is time to hand over the reins to a new generation. This is another area in which Sepp's behaviour has been exemplary. And that's an art. Especially in times like these.

Vienna, June 2017

Ingolf Schädler

Wolf D. Prix

Foreword
A personal speech for Josef Fröhlich that has yet to be held: Complexity need not be complicated.

It is well-known that the older its excellent scientists and artists get, the more Austria likes to shower them with praise and accolades in the hope that they will retire from public life and no longer disturb the tranquillity of convenience with intelligent arguments.

I hope that Josef Fröhlich will not retire from public life. Because…

in older times, one tended to see science and research as a craft enshrouded in secrecy, one that was kept under wraps and used to develop formulas with something resembling sorcery. It took some time to realize – and now this realisation is becoming particularly important – that research and science and maybe also art might be ushering in the future of our development. And to me, this is what Josef Fröhlich stands for as the leader of complexity research in Austria.

With their innovative scientific research, Fröhlich and his team show that we can glimpse the future. It might be blurry, but still.

Sepp Fröhlich has affirmed our feeling that architecture is a closed system, but becomes complex and open when it corresponds to a society that is right for its time. It was his idea that we should develop a building which would give spatial expression and identity to the networked, interdisciplinary research performed by his team. That was in 1991.

I have known Sepp Fröhlich for over 25 years and our conversations about networking and complexity in our society and thus in architecture intensified all the time – and it was clear to us that the influence of his research will have a great influence on architecture.

The knowledge that he conveyed to me has reinforced my reasoning in architectural theory, namely that complex problems can only be met with complex solutions. And that only these solutions – in contrast with simple populist solutions – are innovative.

I remember his lecture at the most progressive architecture school in Los Angeles. The students barely grasped all of its innovative and political implications, but they understood what his remarks were about.

The complex must not be mistaken for the complicated. And I realised that socializing in complex systems requires an overview. Especially in architecture. My friend Sepp Fröhlich explained it to me.

Vienna, June 2017

Wolf D. Prix

Matthias Weber

Introduction

The three terms "Innovation", "Complexity" and "Policy" have been chosen for the title of this volume, because they represent the key concepts that have inspired the work of Josef Fröhlich over more than three decades of research work at AIT Austrian Institute of Technology and its predecessor organisations. Understanding how research and innovation work, how they drive change in the economy, and how government can make sure that socially beneficial impacts can be reaped were the driving force behind his work at the interface between applied research and science-based policy advice. These interests met with a growing recognition by the Austrian government that the country could only maintain its industrial strengths and ensure the standard of living of its citizens by advancing and exploiting the scientific potential of the country.

In line with his background in theoretical physics, Josef Fröhlich was very much inspired by systems thinking, and when system approaches became widely accepted in the economics of innovation and innovation policy alike (Fagerberg et al. 2004; OECD 1992; OECD 2002), his research group took up these ideas very quickly as a foundation to inspire research and innovation policy making in Austria, and also to contribute to the further advancement of innovation systems research (e.g. Schartinger et al. 2002, Weber and Rohracher 2012). Still, in spite of the political success of innovation systems thinking, it represented mainly a heuristic approach rather than a consolidated theory of innovation. The innovation systems approach recognized the evolutionary nature of innovation dynamics, as well as the importance of institutional frameworks in shaping knowledge creation, knowledge exchange and the spread of innovations in the economy and the society, but it did not provide a sufficiently thorough and sound theoretical foundation to satisfy a physicist's heart. However, at about the same time, complex systems science and complexity theory arose as a new scientific promise to better handle not only scientific-technological but also social phenomena. By providing new conceptual and also mathematical instruments it allowed to better capture the nature, patterns and underlying mechanisms of innovation dynamics (Fischer and Fröhlich 2001). The emergence of social and complex network analysis, advanced statistical modelling and new types of simulation techniques like agent-based modelling foreshadowed a new generation of innovation research and science-based policy (see, e.g., Scherngell 2013 for an overview).

But Josef Fröhlich's ambitions went beyond achieving a better scientific understanding of innovation phenomena. He aimed to bring scientific insights to bear on decision making, either in innovation management or, probably even more so, in government policy. This ambition met with the interests of the Austrian ministry in charge of innovation and industrial policy, which happened to be also the owner of AIT. The collaboration with key government officials turned out to be extremely fruitful for the development of an innovation and technology policy, for which by now Austria is recognized as being among the leaders among the industrialized countries of the world. The design of the first Austrian Competence Centres Programme Kplus, which was drawing on studies and concepts of Josef Fröhlich's research team (Ohler et al. 1997), may serve as an important example, just as later work to underpin the re-positioning of Austrian Institute of Technology in the national innovation landscape (Fröhlich et al. 2008). In recent years, these success stories of Austrian research, technology and innovation policy have also attracted much interest outside of Europe.

Another legacy of his work must be seen in the training of a whole generation of specialists, who are now working at international bodies, national ministries, funding agencies, in industry and academia. In fact, the intellectual foundations of several organisations that now make part of the Austrian innovation research landscape can be traced back to the influence of Josef Fröhlich pioneering work. As a lecturer at different universities, and since 2007 as Honorary Professor for Economic Geography at Vienna University of Economics and Business (WU), he has trained and supervised many students. With the Innovation Economics Vienna PhD and Master programme, set up initially in cooperation with the Institute for Economic Geography and GIScience of WU, and now comprising partner institutes from five universities, the efforts to train young researchers in the theory and methods of innovation research became institutionalized up to the present day.

The contributions collected in this volume may not cover the entirety of Josef Fröhlich's work over the past three decades until his retirements in June 2017, but they certainly cover a significant part of it, and probably a part that was very close to his heart. The first Part of this volume brings together a range of contributions which look at the changing patterns of innovation activities and R&I funding in Austria. Rupert Pichler looks from an historical perspective at the changing research funding landscape in Austria, and in particular at the process of agencification of research funding, with all its advantages and drawbacks of first and second order. He observes a very recent development towards "reverse agencification" and points to some major challenges that research funding will need to tackle in the years to come. Doris Schartinger and Bernhard Dachs revisit one of the most influential issues in research and innovation policy around the turn of the millennium, namely the cooperation patterns between industry and academy, and how that picture has changed since then. A second prominent issue for the Austrian innovation systems concerns the international embedding of firms R&D activities. Bernhard Dachs and Georg Zahradnik take a look at recent patterns of foreign-owned business R&D in Austria and the consequences of these investments on the technological capabilities and competitiveness of the country. As a final contribution to this first section, Matthias Weber, Petra Schaper-Rinkel and Susanne Giesecke take a forward look at how R&I activities, organisations and institutions may change in the future. By drawing on some recent foresight work, they present five transformative scenarios that foreshadow major challenges to be addressed by government policy in the years to come.

The second Part concentrates on the methodological dimension of innovation research, and in particular at advanced methods to capture in quantitative terms how research and innovation activities change in thematic and structural terms. Thomas Scherngell, Laurent Bergé and Iris Wanzenböck show by way of the example of bridging centrality how Social Network Analysis can contribute to the development of new meaningful positioning indicators that reflect the knowledge bridging role of individual organisations in a network of R&D collaborations. Capturing the complex dynamics of innovation networks is also the ambition behind agent-based modelling. Manfred Paier, Martina Dünser, Thomas Scherngell and Simon Martin demonstrate the potential of agent-based modelling by looking at the influence of different types of policy interventions, either stimulating or discouraging risk-taking, on innovation performance. Edgar Schiebel focuses in his contribution on the thematic dimension of research funding, by using bibliometric tools to identify research fronts in the field of security research.

In the third Part, the emphasis is put on the analysis of interventions in innovation systems. In the contribution by Michael Dinges and Karl-Heinz Leitner, the Austrian competence centre programmes are interpreted as systemic instruments that exert an influence on innovation performance through a whole range of mechanisms. Eva Buchinger focuses on demand-side innovation policy, and more specifically on the development of public procurement as a new instrument of innovation policy, thus complementing the more established supply-side instrument of R&I funding. In line with the idea that new forms of intervention are needed to influence complex social systems. Barbara Streicher, Margit Fischer, Wolfgang Czerny and Barbara Heller-Schuh present the example of the Austrian Science Center Network, which has gained a lot of momentum since its foundation in 2005 and managed to create a new level-playing field in the area of science learning in Austria. The growing influence on innovation of users and their experiences is at the core of Manfred Tscheligi's contribution. This new role of users has long been under-estimated in management as well as government policy, but is acquiring growing recognition due to the widespread use of IT as main interfaces between humans and machines. The final contribution to this volume, by Ludwig Streit, looks into fundamental issues of the difficulties of intervening in complex social systems, and the implications this raises for the effectiveness and credibility of democracy.

References

Fagerberg, J., Mowery, D.C., Nelson, R.R. (eds.) (2004). The Oxford Handbook of Innovation, Oxford: Oxford University Press.

Fischer, M.M., Fröhlich, J. (eds.) (2001). Knowledge, Complexity, and Innovation Systems, Berlin, Springer.

Fröhlich J., Weber M., Brückl H., Dachs B., Garn H., Haberhauer G., Kubeczko K. (2008). Perspektiven für die außeruniversitäre Forschung in Österreich: Das Beispiel ARC, Austrian Research Centers GmbH – ARC, ARC-sys-0160.

OECD (1992). Technology and the Economy: The Key Relationships, Paris, Organisation for Economic Cooperation and Development.

OECD (2002). Dynamising National Innovation Systems, Paris, Organisation for Economic Cooperation and Development.

Ohler, F., Sturn, D., Stampfer, M., Warta, K., Fritz, O., Fröhlich, J. (1997). Kplus – Forschungskompetenz plus Wirtschaftskompetenz. Vorhabensbericht des zur Errichtung von Kompetenzzentren in Österreich, Vienna, Federal Ministry for Science and Transport.

Schartinger, D., Rammer, C., Fischer, M. and Fröhlich, J. (2002). Knowledge Interactions between Universities and Industry in Austria: Sectoral Patterns and Determinants. Research Policy, 31(3), 303-28.

Scherngell, T. (ed.) (2013). The Geography of Networks and R&D Collaborations, Berlin, Springer.

Weber M., Rohracher H. (2012). Legitimizing research, technology and innovation policies for transformative Change. Combining insights from innovation systems and multi-level perspective in a comprehensive 'failures' framework, Research Policy, Vol. 41, 1037–1047.

Acknowledgements

The preparation of this volume would not have been possible without the continuous and thorough support by Beatrice Fröhlich-Rath.

The dynamics of innovation systems

Rupert Pichler

RTI funding in Austria – Organisational structures and institutional change

Abstract: *Austria has a mature RTI funding system today. Being a late-comer in terms of institutionalisation, Austria developed councils and agencies as the standard organisational models since the 1960s. Institutional path dependencies, fragmentation, and contradicting claims of autonomy and political steering led to persistent tensions. Recent changes were therefore driven by the need for better coordination and rebalancing principal-agent relations.*

1 Introduction

Today, Austria has a well-developed funding system covering the whole spectrum of research, technology and innovation (RTI). Within the institutional framework of RTI policies, the implementation of RTI funding usually requires a specific organisational approach with structures that mediate between government and the RTI community. Accordingly, the presence of RTI funding organisations (RFOs) indicates the maturity of a national innovation system: "Funding agencies were, since their origins, designed to work out and implement research policies, in preference to the usual public bureaucracy that lacked the necessary direct contacts with science." (Braun and Guston 2003, 303) On the contrary, it can be observed that substantial direct government funding only survives in institutionally less developed systems (Lepori et al. 2007, 385).

At federal level, there are two big organisations which entirely or in part are dedicated to the implementation of RTI funding programmes: the Science Fund (FWF) and the Austrian Research Funding [official translation: Promotion] Agency (FFG). In addition, some innovation-oriented programmes are assigned to the Austrian Economic Service (AWS). In 2015 FWF approved funding worth 199.3 million Euro, FFG 467.1 million Euro (including loans and liabilities). AWS awards most of its 825.6 million Euro (2015) as loans and liabilities[1] (BMWFW and BMVIT 2016, 89, 94, 98). The sources of these funds are largely the federal budget, but also (public) foundations as well the provinces. The latter also dispose of RFOs themselves. These, however, exceed the scope of this paper.

1 By far, the largest part of the AWS funding does not address RTI.

Beyond RFOs, RTI *performing* organisations (RTOs) constitute an important element of innovation systems. There are some big Austrian players like the Austrian Academy of Sciences and the Austrian Institute of Technology, but the tradition of governmental research organisations is rather weak compared to other countries (Barlösius 2010; Bach and Döhler 2012; Bach et al. 2013). While RTOs are in relation to RFOs potential applicants and subjects of funding, there are also hybrid structures. Throughout the institutional development of research policies we can observe organisations which combine the tasks of funding and performing research or are research organisations having internal mechanisms in place for the allocation of resources (Braun 1997). The former was, for example, the case with the British research councils until the advent of the Higher Education Funding Councils. Obvious examples for the latter can be found in Germany (Hohn and Schimank 1990; Szöllösi-Janze 1990; Trischler and vom Bruch 1999; Groß and Arnold 2007; Hohn 2010), including communist experiences (Gläser/ Meske 1996). In Austria, this model is represented to a certain degree by the Christian Doppler Association and the Ludwig Boltzmann Association, though on a much smaller scale.

It is, therefore, the "classic" RFO structure providing the dominant organisations for RTI policy implementation. By the same token, RFOs are at the centre of institutional change and innovation in Austria's RTI policy which benefitted for more than 30 years from Sepp Fröhlich's and his department's expertise. This paper tries to analyse the nature of these structures and recent changes leading up to the present situation.[2]

2 Institutional framework of RTI funding

2.1 Agencies and councils: a typology of RFOs

RFOs are often termed agencies. Agencies are an essential element out of the New Public Management (NPM) tool box. RFOs as funding agencies also count among the most important intermediaries in the field of research policy. In the NPM context agencies are aimed at ensuring better goal attainment through greater efficiency. Basically, the latter is achieved by the division of labour and clarification of roles as according to the NPM logic agencification means separating regulatory from operational tasks (Christensen and Lægreid 2006; Talbot 2004; Van Thiel 2012; Bach and Huber 2012).

2 Some parts of this paper have already been published (Pichler 2016).

One underlying model is the principal-agent approach: the principal delegates a task to the agent who knows more than the principal about a specific field where the principal wishes to intervene (Braun and Guston 2003). Such a relationship is typical for the implementation of RTI policies which "needs the co-operation of scientists to overcome the implicit lack of knowledge or, in other terms, the 'information asymmetry' inherent in functional differentiation." (Braun 2003, 310)

In that sense, RFOs are agencies. But as their primary aim is not increasing efficiency and effectiveness, this is also misleading. Most RFOs had been established long before the NPM wave began to roll. Beyond NPM and its neglect of the historical context, there is a long tradition of "third sector bodies which were originally voluntaristically created to pursue the goals of their membership, but which subsequently became responsible for the delivery of certain public services." (Pollitt et al. 2005, 9) In Austria and Germany, for example, sickness funds, Chambers of Commerce and the like follow that pattern so this was the model at hand when RFOs were first discussed.

There is yet another aspect of the specific nature of RFOs as agencies. Much of the NPM style agencification model is about regulatory reform. But RTI policies are distributive rather than regulatory policies: "Distributive policies ... have low visibility, are non-conflictive, and have a low salience in policy making. This is easily explained: as there are no visible costs involved, we seldom find coalition building to defend particular issues in this context. Mostly small and homogeneous groups profit from the public money." (Braun 2006, 148) As a result, low salience and homogeneity of the beneficiaries may considerably increase the risk of agency capture.[3]

Trying to define whether or not RFOs are agencies obviously requires criteria so as to distinguish agencies from other intermediary RFOs. Generally, an RFO can be defined "as a non-ministerial institutional structure in the public domain designed to manage the allocation of government funds to research projects, individuals and/or institutions. Its governance structures enjoy considerable autonomy from the government and integrate those addressed (e.g. the scientific community) into its decision-making" (Stampfer et al., 2010, 766). The concept of agencification fits into this definition, but it is not the whole story, not least because most RFOs had been established long before NPM was introduced.

Beyond the reasons *why* the implementation of certain tasks is separated from ministries, a differentiation can therefore be made in functional terms along the

3 Agency capture is a failure where government agencies begin to be controlled by those they address (Braun 1997, 56–59).

modes of delegation – delegation by trust versus delegation by contract (Braun, 1997 and 2003). Two basic categories of RFOs can be derived from that (Stampfer et al., 2010, 767):

- The council, which manages its funds autonomously and is run by members of the communities that it addresses.
- The agency, which provides the government with managerial expertise and capacities and manages its funds on the government's behalf.

These categories translate into different ways how council- and agency-type RFOs meet their intermediary function vis-à-vis the RTI community. Table 1 defines criteria for that function and specifies the different characteristics.

Table 1: comparison of intermediary functions

intermediary function	**council**	**agency**
legitimation	corporative	technocratic
relation to government	trustee	service provider
role of those addressed	members	customers
role of governing bodies	interest grouping	provider of expertise
expected benefit	shared responsibility	division of labour

Source: Pichler 2016, 36.

It becomes fairly clear that these two models can be well differentiated, not ignoring that in some administrative cultures, the organisational set-ups (like e.g. in France) are dwelling between these two strands (Groß et al. 2010). Even so, it is incorrect that in Austria the RFOs are commonly just referred to as "the agencies". It has also to be mentioned that historically they may represent subsequent stages of institutionalising research funding. Looking at FFG, we will see the relevance of this in historical perspective.

Furthermore, this distinction reflects different degrees of autonomy. It is usually with basic research that the council model persists whereas applied research is often subject to closer political control, because "different degrees of independence or autonomy are strongly linked to uncertainties of policy makers regarding the outcomes of the action of scientists. If there is high uncertainty and the likely gains of research investments are in the distant future, delegated agencies may have both substantial discretion and be exempt from political control. The more visible and likely research results become the more policy makers attempt to reduce discretion and exercise control over funding agencies." (Braun 2006, 167)

Arguably, it is the conceivable opportunity to appropriate research results that determines how RFOs are shaped in terms of autonomy. However, because RFOs are always somehow or other connected to their stakeholders they usually enjoy "high autonomy in terms of policy discretion" (Bach 2016, 2).

2.2 Legal framework

Notwithstanding the widespread applicability of the RFO model as well as agencification in other policy fields, different legal frameworks account for a broad variety of modes in which RFOs are implemented that are difficult to compare.[4] In international comparison, the function of RFOs can be achieved with or without legal personality, in public or private law, using tailor-made or standardised models or existing organisations (Van Thiel 2012; for RTI, see Braun 1997, fig. 2 and 3). Examples for standardised public bodies are the non-departmental public body in the UK, the etablissement public à caractère scientifique et technologique in France (both having full legal capacity) or the independent federal agency in the US (without legal personality). As private law models associations, foundations or publicly owned companies are used. The utilisation of existing organisations not originally set up as RFOs is also an option. Also the EU embarked on agencification in research funding, the most prominent example being the European Research Council with its executive agency (König 2016). Depending on the respective model, steering is exerted by oversight, instruction and/or contract.

Austria, not unlike Germany, represents a system with a strongly legalistic administrative culture. Hence, bureaucratic action is legally bound and executed in a chain-of-command manner: "Hierarchy thus becomes an almost sacred principle because it is seen as a safeguard for executing the law as intended by parliament. … This doctrine is complemented by the principle of ministerial responsibility", where "each deviation from the hierarchical model … requires thorough substantiation." (Bach and Döhler 2012, 10)

The scope of agencification in Austria is therefore limited, not least because of the constitutional court's rulings which almost completely preclude the delegation of any regulatory matter.[5] RTI funding and financing, however, is no such matter in which case there is more room for manoeuvre as the federal constitution grants considerable freedom for any action based on private law (article 17). As a result, it

4 Efforts to do so have been undertaken nonetheless, as for instance with the "Governance Equalizer" (de Boer et al. 2007).

5 Where there are regulatory agencies they are in fact only the administrative bodies of otherwise "classic" public authorities.

is feasible to create RFOs as separate legal entities provided they award their funds on a contractual basis (i.e. under private law). For this purpose, tailor-made public (statutory) bodies (no standardised models exist), publicly owned companies, foundations or associations may be established or existing organisations may be entrusted with that task (Pichler and Strimitzer 2013). Within these limits, such entities can be endowed with a high degree of formal, decision-making autonomy.

One notable exception though to those rules is the tax credit for R&D which is subject to tax regulations and which is administrated by the tax authorities. But here the aforementioned lack of knowledge of the bureaucracy became visible over time, so that FFG was mandated to supply its expertise in the course of the approval procedures (BGBl. I Nr. 22/2012, § 108c EStG). For the establishment and/or funding of RTOs basically the same principles as for RFOs apply. Yet there is also a tradition of RTI capacities inside the administration (as "Bundesanstalt") – though more closely connected to academia – where specific knowledge was needed for regulatory tasks, like in standardisation or health policies (Pichler and Hofer 2014, 21–22).

2.3 History of RFOs in Austria

Compared to other European countries, it took Austria a long time before RFOs were first established. Not only was any pre-World War II tradition of such organisations lacking, the specific political set-up of post-war Austria greatly contributed to an almost endless hassle about the institutionalisation of RTI funding owed to the overarching principle of the division of power between Christian (ÖVP) and Social (SPÖ) democrats (for the whole story see Pichler et al. 2007, Stampfer et al. 2010, Pichler and Hofer 2014).

Beyond the mere power play, the two parties pursued different goals with their proposals. The ÖVP wanted to strengthen existing academic institutions, primarily the universities and the Academy of sciences to which it was traditionally closely linked. The SPÖ, on the other hand, whose ties to the academic world were weak, preferred a type of RFO which would lead to new, additional research institutes reaching out also to industry and society. As a consequence, these different goals led to different institutional preferences. The ÖVP proposals envisaged a council-type self-administration by the scientific community, the SPÖ vision included stronger political guidance (of a council-type structure nevertheless).

As long as those two parties found themselves in a coalition government their concepts could not be reconciled so as to eventually lead to a legal provision for an RFO. Instead, both of them tried to create a fait accompli by supporting the foundation of associations designed to function as RFOs, namely the ÖVP-bred

Österreichischer Forschungsrat and the SPÖ-bred *Ludwig Boltzmann-Gesellschaft*. Their bylaws markedly reflected the priorities of the respective political godfathers. However, these were small games about small sums of money so that in the absence of a compromise on an RFO, bigger investments were directed to RTOs in each of the parties' domain, like the predecessors of AIT.

Only in 1966, when the grand coalition system broke up for the first time, the door was opened for a law on RTI funding. But now the ÖVP could design everything to its own choice, not meaning that this was a quick win because within the ÖVP sphere industrial interests had strongly voiced their claim on research funding. This resulted in the creation of two RFOs, one for academic and industrial research each. Accordingly, the Research Funding Act of 1967 provided for two public bodies called funds, FWF and FFF. Their organisational set-up followed the ÖVP principle of strong autonomy and self-administration where the government would not even appoint board members. In international comparison FFF was a very peculiar institution as for the funding of applied research the council model – which put FFF largely into the hands of the chamber of commerce as "appropriate" representative of the community addressed – was highly unusual. Given the higher visibility of applied, company-driven research one would have expected a stronger political will to control (Braun 2006, 167). Yet another ÖVP principle was carved in stone with FWF and FFF, that is the project-only philosophy which ruled out any kind of institution-building within the funds' realm.

FWF (which also succeeded the *Forschungsrat*) and FFF soon became highly recognised within their respective communities. When in 1970 the political pendulum swung in the opposite direction and the SPÖ took over, it was thought unwise to change the funds along the own preferences. Rather, the first-time introduction of a ministry of science and research was intended for political steering. In order to make up for what the funds did not fund – mainly top-down set priorities – the ministry also developed its own in-house funding facility. This was a strange step at a time when Austria had just managed to create state-of-the-art RFOs. But soon, the ministry's funding encountered expectable troubles because of weak feed-back loops to the scientific community and resulting low credibility there.

This lesson learnt, a different path was chosen in order to cater for more complex, often cooperative programmes and projects when technology policy entered the scene. A new fund, the Innovation and Technology Fund (ITF), was installed and managed by FFF together with the ERP fund in an agency-like manner (Biegelbauer 2013, 75–144). This was complemented by the foundation of an association together with (at the beginning: state-owned) industry, the *Christian*

Doppler Gesellschaft (CDG) which funds joint industry-university labs, given the growing importance of university-industry cooperation (Pichler and Hofer 2016). Another massive boost came with the accession to the EU when Austria got to know the framework programme's RTI funding world based on strategic programming, competitive calls, cooperation, and evaluation. These approaches were soon applied to new, ambitious programmes in Austria like the Competence Centres (Biegelbauer 2007; Biegelbauer 2013, 145–172) which, however, required strategic expertise and appropriate implementation structures. The strategy paper for the competence centre programme "K plus" was written by Sepp Fröhlich's department at AIT, then Austrian Research Centres Seibersdorf (BMWV 1997). For the implementation of, amongst other programmes, "K plus" a new agency, the Technologie Impulse Gesellschaft (TIG) was established, but the Austrian Space Agency (ASA) also took over the management of some programmes. Table 2 shows the "animals in the administrative zoo" (Bach and Jann 2010) of RTI funding in Austria whose latest addition was going to be FFG.

Table 2: development and characteristics of RFOs in Austria

Organisation	lifetime	legal form	functional type	kind of research
Forschungsrat	1960–1967	association	council	basic
LBG	1960--	association	council	basic
FWF	1967--	public body	council	basic
FFF	1967–2004	public body	council	applied
ASA	1972–2004	limited company	agency	applied
ITF	1987–2000	public body	agency*	applied
CDG	1988--	association	council	industrial basic
TIG	1998–2004	limited company	agency	applied
FFG	2004--	limited company	agency	applied

Source: Pichler 2014, 336.
* managed by FFF and the ERP-fund

3 Agencification and "reverse" agencification

3.1 Structural reform and the foundation of FFG

By the turn of the century, the administrative zoo was well-populated with animals of different age, size, shape, experience, and needs. Contrary to a real zoo, diversity was not a goal. A widespread discussion led to a substantial evaluation whose subjects were FWF and FFF. The starting point of the debate leading to the evaluation was the fragmentation and multiplicity of the funding system. But soon the autonomy of FWF and FFF got into the focus as an element of petrification and lacking accountability: it seemed obvious that "most of the innovation in Austrian innovation funding system has taken place outside FFF" (Arnold 2004, 57). The reason for that was that the funds "were given a strongly 'autonomous' status, which is to say that they were given governance structures that were dominated by their beneficiaries, rather than by ministries. This reflected a lack of confidence that they could be kept free from inappropriate, detail-level interference … What the Research Promotion Act set out to do through autonomy, is elsewhere being achieved through increased clarity about the respective roles of ministries and agencies" (Arnold 2004, ii). Eventually, this "proved to be a tragic flaw, because to a considerable extent it locked the Funds into their 1967 roles" (Arnold 2004, 102).

Derived from this analysis the evaluation report recommended that "the balance of power in its governance structure should be changed so that more representatives of the taxpayers and not the beneficiaries have control" (Arnold 2004, 59, on FFF). As a result, the funds "need to be changed from quasi-autonomous bodies to agencies of the government" (Arnold 2004, 113). Or, not in so many words: the funds were stricken by agency capture.

As a consequence, when FFG was created in 2004 this was the result of a reform whose narrative was much about NPM-style clarification of roles. Therefore, the foundation of FFG was widely conceived as agencification. Yet by 2004 the structure that was now chosen for FFG had already been operated successfully by one of FFG's predecessors (TIG), based on a more timely delegation-by-contract approach. Other attempts, though, to install elements of agencifcation within FFF and FWF had been doomed. When in the late 1990s many new programmes had to be implemented, the ministry needed the operational expertise of FFF where TIG could not be used because of its specific focus. Thus the legal provisions for FFF were amended by a regulation adding the implementation of contracted programmes to its remit (BGBl. I Nr. 79/1999). However, the governance structures of FFF remained unchanged, thus keeping their responsiveness towards ministerial intentions at a low level. The new task just seemed to be a nice add-on, leaving the general approach of FFF unchanged. FWF got into a similar situation, but as the programmes in question had been born out of FWF's own community, a legal amendment was not deemed necessary for the time being.[6]

6 Nonetheless, this was made up for in the course of the 2004 reform bill.

Within this context of gradual, yet uncoordinated institutional change – supplemented by the introduction of the council for research and technology development as a new advisory body to the government – the evaluation of the funds was among the triggers of efforts to change the legal set-up of research funding substantially. The debate soon concentrated on what was to become FFG. The evaluation strongly advised to leave FWF out of a merger of FFF with other RFOs which was indeed recommended (Arnold 2004, 13–17, 107) while the Court of Auditors – just having examined the funds too – said the opposite (Rechnungshof, 2003). Moreover, the strong stakeholder-driven tradition of FWF as a council-type RFO delayed any agency-like change of structures.

In a nutshell, the idea of agencification was relevant within the context of the evaluation of FFF and FWF, and following that for the creation of FFG. But the major point was what can be called "reverse" agencification, a process that had taken place before in some form in other countries too. Already in the 1960s the research councils in the UK had come under pressure (Braun 1997, 194–199, 298–302) and the claims of the German research ministry (Stucke 1993) had been fiercely disputed by DFG (Orth 2011, 100–116).

Despite a certain, and sometimes dodgy, tradition of ministerial in-house funding, its delegation did not cause much debate at the time of the beginning of the 2000s anymore, not least because of the ministries' staff restrictions. While this process – entailing e.g. the foundation of TIG 1998 – was well underway, the main problem that had been addressed by the evaluation was too much rather than too little independence of research funding. Therefore, agencification proved to be a reference for reducing instead of increasing autonomy.

3.2 The institutional set-up of FFG: parallel worlds surviving

Taking the recommendations of the 2004 evaluation further, the obvious question is whether or not they were implemented. Firstly, the research funding reform bill of 2004 (Forschungsförderungs-Strukturreformgesetz, BGBl. I Nr. 73/2004) proposed a merger of FFF with BIT (an association advising on EU framework programme participation), TIG and ASA. The new agency (FFG) took the form of a state-owned limited liability company. Contrary to FFF, however, the management of FFG is thus bound by instructions of the minister(s) in charge. The explanatory note accompanying the government's proposal features many NPM buzzwords: it called for modern governance structures, a clear attribution of responsibilities, harmonisation of fragmented structures, less transaction between ministries and intermediary bodies, all supposed to lead to simpler structures creating synergy effects and more transparency (Pichler 2014, 37).

The choice of a limited liability company was standard procedure. The consequences of the principals' ownership in terms of their influence were intended. In that sense, the government did what the evaluation had called for, including respective regulations about the balance of power in the governance structures. The board of FFG has now a majority of members appointed by the ministries (excluding the employees' representatives), compared to the non-voting position of their (few) representatives in FFF times. In the same vein, the proposed merger of FFF into a broader agency was followed up, explicitly justified by the expected gain in efficiency during the legislative procedure.

As the creation of FFG was not an act of agencification from scratch, a closer look reveals what effects the legacy of its predecessors had on its development as an agency. As a funding agency, the principal task delegated to FFG is to implement the spending of public funds. However, what the 2004 law regulates in the first place was the delegation of management responsibilities. It says little about the delegation of financial responsibilities. The level of management delegation results from the principles guiding a limited liability company, as described above. As regards the allocation of funds things are less clear. The law only suggests that FFG is to allocate funding on its own as well as on the government's behalf. In legal terms, this principle is only fully understood together with other legal provisions governing research funding (e.g. Forschungs- und Technologieförderungsgesetz). Hence, different modes from the past are surviving: FFF was autonomous not only in management, but also in financial terms as it decided on the distribution of its funds; both TIG and ASA were limited liability companies, but TIG got injections of capital while ASA handled the money only on behalf of the government (like FFF when it was also assigned to manage additional programmes).

FFG inherited parts of these different approaches. FFG disposes of funds underlying its own allocation decision as well as funds where it only executes the respective ministry's (formal) allocation decision. This puts the principals in a dual role as both owner and client. This dichotomy also feeds back to the delegation of management tasks as the different levels of financial responsibility require different ways of planning and reporting.

Summing up, there is some evidence of incomplete agencification: while the autonomy of FFF was reversed to the level of a state-owned company with decision-making power on the allocation of funding, FFG was not empowered to manage all the other programmes and respective funds on that very level. On the other hand, in-house ministerial programme implementation had almost ceased to exist after FFG came into being, which was a crucial step in the process of mainstreaming the government's approach towards the delegation of RTI funding.

3.3 Agencification in post-2004 developments

Neither the law establishing FFG nor any other legal provision at that time required that FFG be evaluated after a certain period of time. Most evaluations carried out in these years largely focused on programmes and their impact. The institutional perspective of research funding was featured again in the system evaluation of 2009 (Aiginger et al.). Even though this evaluation primarily addressed the effect of the research funding system on the target groups' behaviour, the underlying structures were also looked at.

Quite in line with what was said above the evaluation stated: "Agencification has been started but has not been fully implemented" (Aiginger et al. 2009, 140). This seems to have been judged as a flaw compromising the division of labour between ministries and FFG, amongst others "caused by the existence of two parallel steering mechanisms" (Aiginger et al. 2009, 86). After all, the owner-client dichotomy was not lost on the system evaluation. It proposed that the principal's influence on the agency should generally be exerted through its right of ownership and a performance agreement. Through "a process of earned autonomy" this would lead "to more operational freedom of the agencies and more room for strategic governance at the side of the principal" (Aiginger et al. 2009, 141). However, "that should be based on a prior normative discussion about the positions on and understanding of the division of labour, governance and STI policies as such" (Aiginger et al., 2009, 88).

After the structural reform the issues of autonomy and delegation surfaced again in connection with FWF. After 2004 the traditional decision-making structure of FWF had essentially remained unaltered (except for the introduction of a supervisory board), meaning that the financial responsibilities had stayed with the president whose principal legitimation, however, is based on academic merit. In recent years, this led FWF into trouble with the new public corporate governance and budgetary requirements, as well as with the court of auditors (Rechnungshof 2015).

As a consequence, an amendment of the legal provisions for FWF was proposed. What the corresponding impact assessment states can be – once again – interpreted as reverse agencification, at least in part (Pichler 2016, 38). Firstly, it was made clear that in legal terms FWF is the result of delegation and thus (almost) an agency. Taking into account the rulings of the constitutional court on the delegation of public tasks the government had therefore to regain more influence on the governance of FWF. Secondly, the division of responsibilities between president and management was to be clarified, and the competences of the supervisory board were to be extended in the way applying to limited liability companies. As a result, FWF got governance structures much closer to an agency (BGBl. I 110/2015). It now practically has a managing board, including a CFO on equal terms with the scientific president, reporting to the supervisory board. The president and the CFO are appointed by the board, the former based on a proposal by the assembly of delegates, instead of the other way round before. This fulfils the constitutional requirements as the board includes members appointed by the government. In short, the old council days are, to a certain extent, gone. Where accountability meets autonomy, clarity on responsibilities and the return for the government beats the old glory of academic self-administration. Here again, there is some sort of "reverse" agencification.

Starting from the mid-1990s, Austria has witnessed changes in its RTI funding structures that can be characterised as – or were decidedly meant to be – agencification. One key promise pertinent to NPM-style agencification is clarity and simplicity to allow for better steering. In other words, complexity counts as a functional problem as such. As regards the number of agencies, FFG reduced complexity. Although one might argue that smaller agencies could be steered better and that a certain diversity may even prove advantageous[7], the key point in 2004 was the "increased clarity about the respective roles of ministries and agencies" (Arnold 2004, ii).

It may be questioned to which extent this has, in fact, been achieved. The persistence of different delegation levels hints at some problems with the separation of strategic and operational tasks. Hence, the checks and balances between agent and principal may be jeopardised. The balance "is more likely to be stable ... when both need the other" (Pollitt et al., 2005, 23). Since this mutual dependence can turn out differently according to the respective delegation level, as a result the principal(s) may alternate between too much and too little guidance of the agency.

The jury is still out on how much – if at all – this affects the performance of FFG. Despite the somehow twisted design of its formal autonomy level, the strength and *actual* autonomy of FFG lies in its "ability to cultivate a coherent mission and legitimacy among stakeholders" (Bach 2016, 1). The ongoing institutional evaluation of FFG and AWS whose results are expected by mid-2017 may cast some light on these issues and eventually pave the way for yet another step towards institutional change.

---

7 As often argued in the case of the German *Projektträger*.

4 RFOs and RTI policy coordination

In theory, the advantage of agencies lies in the division of labour and clear attribution of tasks. Thus, also in the field of RTI "intermediary agencies are mono-organisations, i.e. they are focusing on one sector only, while ministries often entail more than one sector, which contributes to a fragmentation of powers and objectives", and at the same time "agencies are often more closely linked to their clientele at the operative level than ministries" (Braun 2008a, 235). Such a constellation clearly leads to a situation where agencies are an obstacle to better coordination, even more so within RTI policies as they are by nature cross-cutting rather than sectoral with strong vertical linkages: "New public management delegation is obviously not built to develop networks and cooperation … It is developed to make vertical relationships more efficient but not horizontal ones." (Braun 2008b, 294)

But still "delegation is not conceived as a means for coordination but it can of course be used for this purpose." (Braun 2008b, 293) In the Austrian case this emerges from some of the rationales for the creation of FFG which was to be used to come to terms with the tension between the horizontal nature of RTI and the intrinsic role of RFOs in the implementation of RTI policies. Even though in line with the FFF/FWF evaluation and contrary to the court of auditors' recommendation a merger of FFF and FWF war renounced, the fragmentation of the RFOs was an explicit reason for FFG taking the form of a merger (Pichler 2014, 339). Already the 2004 evaluation had found "a surprisingly high degree of fragmentation among its innovation agencies" in Austria and suggested: "FFF should be merged into a broader innovation agency." (Arnold 2004, 107, 114) CDG and LBG, however, were excluded from these considerations as they are hybrids between RTI funding and performing and would thus not fit well into the horizontal coordination capacities of FFG.

As a result, FFG is a multi-principal agency serving a number of ministries (its two owners, but also other ones as clients) as well as provincial governments and the EU today, reflecting the fact that RTI is an element within the realm of multiple ministries rather than a sector of its own. Certainly, it is not the potential closeness of FFG to its clientele that may hamper its use for RTI policy coordination, since the undue influence of the beneficiaries on "their" agency is not a problem anymore, neither with FFG[8] nor – for the better part – with FWF, but the risk of "reverse" agency capture (or "ministry capture") which cannot be fully excluded. One possible reason in the case of FFG is its very nature

8 That FFG terms its applicants „customers" is owed to NPM wisdom, rather than an imminent danger of agency capture.

as a multi-principal agency, potentially deepening the information asymmetry between FFG and its principals.

That is because FFG is put in a position where it has a better overview and thus more information than individual ministries, all the more as it manages to attract more ministries as clients. For instance, since FFG manages the review of the tax credit applications on behalf of the finance ministry, it is the only one having an overview of the applicants' activities in both competitive and tax credit programmes. The law governing FFG even enhances this effect as it puts FFG in a position to propose strategic planning documents that have to be approved jointly by the owner ministries, so that they are coordinated *by* the agency rather than exerting coordination themselves *through* the agency.

Even if this yields positive coordination effects (Pichler 2014) it comes at a price. Limited coordination capacities at government level have allowed FFG to fill this strategic gap without being able to entirely bridge it as an agency (as a general notion expressed by Braun, 2008b, 298). Once again, the consequences are – contrary to what was intended – blurred boundaries between the principals' and the agency's tasks.

Yet Austria is not the only example for such an approach. The Research Council Norway (RCN) actually covers a far more extensive scope than FFG does, serving a much larger number of ministries. It is the only organisation of that kind. After a troubled start in 1993, it seems to have helped break up sectoral silos and gain more weight for RTI as an overall policy priority (Skoie 2000; Arnold and Mahieu 2012).

5 Drivers of change and future challenges

Only a short stretch of time in the development could be looked at in greater detail in this paper. But then it was a period when much of the initial institutional design changed eventually since the 1960s. This was the harvest of what was sown in the 1990s. Throughout this decade, Sepp Fröhlich's department was part of the influential Technology Information and Policy Advice Programme (TIP), which supplied much strategic intelligence to RTI policies at that time. Summing up, we try and identify some of the factors that led to institutional change during the development of RFOs in Austria:

- *The need for competitive funding:* Despite all dissent over the precise nature and scope of an RFO, it was clear to everyone involved that some kind of additional, merit-based funding was necessary. Indeed, the first introduction of RFOs freed innovation in universities and companies from hierarchical restraints.

- *The need for technology funding:* When technology policy became an issue, the traditional bottom-up approach of RTI funding was not sufficient anymore. Addressing technological issues involving multiple actors required structures allowing for stronger leadership.
- *The need for cooperation funding:* The increasing importance of science-industry relations and the funding of projects carried out jointly by academic and industrial players required structures whose expertise covered both basic as well as applied research.
- *The need for strategic programming:* Ever more complex issues and projects, together with the rapid internationalisation following the accession to the EU brought RTI policy back into the driving seat. Therefore, funding structures needed to adapt to a certain degree of political guidance.
- *The need for coordinated funding:* Meeting all these challenges had led to a fragmented set of RFOs with different modes of operating. Achieving a coherent policy implementation required the harmonisation of at least some of these organisations.
- *The future need to balance competition and cooperation:* RFOs count among the most eminent RTI policy players in Austria. Most of the institutional innovation in RTI policy developed in connection with RFOs. Hence, other structures remain underrepresented compared to other mature innovation systems. Particularly larger networks of institutes like Max Planck or Fraunhofer are missing, as the bigger Austrian equivalents were rather designed as temporary structures. In the long run, competition as a predominant governance mode may weaken the overall stability of the innovation system and hamper the division of labour within.
- *The future need for a coherent and comprehensive steering logic:* Today, FFG is governed by both owner instructions and contracts, resulting in multiple, parallel relationships between FFG and its principals as well as other clients. In order to fully exploit FFG's potential this structure should be reconsidered, but also broadening its scope towards stronger impact orientation, and perhaps new instruments may be part of these considerations which to a certain extent are taken into account by the ongoing evaluation.

6 Conclusions

The development of RTI funding in Austria exemplifies the importance of RFOs within the differentiation of an innovation system and the path-dependence creating capacity the initial institutional design can have (Lepori et al. 2007, 376). The latter is caused by the horizontal and expertise-dependent nature of RTI policy requiring the consensus of many stakeholders[9] which made it difficult to first launch the institutionalisation of RTI funding and change these structures, once established, later on. Therefore, the short-term way of institutional innovation was adding new elements to the system (or bypassing it) until their parallel operations alongside each other would give way to a new long-term solution like FFG. A common element of those innovations was the readjustment of the principal-agent relationship according to emerging challenges. As obviously this relationship cannot be kept permanently stable, there are "lasting tensions" (Braun 2003) oscillating between agency capture and government interference. However, if RTI funding gets close to one of these points, the structures get dysfunctional as the examples of FWF and FFF on the one hand, and in-house ministry funding and the tax credit administration on the other hand show. Therefore, capacities of institutions within an innovation system to adapt to changing needs are crucial, even more so in Austria, given the dominance of RFOs over RTOs. This has saved Austria in the past from massive path dependencies like in other countries (Wieland 2009), but calls for readiness in the future.

References

Aiginger, K., Falk, R., Reinstaller, A. (2009). Reaching out to the Future Needs Radical Change. Towards a New Policy for Innovation, Science and Technology in Austria. Synthesis Report. Wien: WIFO.

Arnold, E. (2004). Evaluation of the Austrian Industrial Research Promotion Fund (FFF) and the Austrian Science Fund (FWF). Synthesis Report. Brighton/Wien: Technopolis.

Arnold, E., Mahieu, B. (2012). A Good Council? Evaluation of the Research Council of Norway. Synthesis Report. Brighton: Technopolis Group.

Bach, T. (2016). Administrative Autonomy of Public Organisations. In: A. Farazmand (ed.) Global Encyclopedia of Public Administration, Public Policy and Governance. Springer International, 1–9.

Bach, T., Döhler, M. (2012). Mandated science and the problem of neutral expertise. The case of governmental research agencies. Berlin: Wissenschaftszentrum Berlin für Sozialforschung.

Bach, T., Huber, E. (2012). Comparing Agencification in Continental Countries. In: K. Verhoest, S. van Thiel, G. Bouckaert, P. Lægreid (eds.) Government Agencies: Practices and Lessons from 30 Countries. Basingstoke: Palgrave Macmillan, 203–210.

9 Kritzinger and Pülzl 2008 contest this view for the case of FFG.

Bach, T., Jann, W. (2010). Animals in the administrative zoo: Organisational change and agency autonomy in Germany. International Review of Administrative Sciences 76(3), 443–468.

Bach, T., Philipps, A., Barlösius, E., Döhler, M. (2013). Governance von Ressortforschungseinrichtungen. In: E. Grande, D. Jansen, O. Jarren, A. Rip, U. Schimank, P. Weingart (eds.) Neue Governance der Wissenschaft. Reorganisation – externe Anforderungen – Medialisierung. Bielefeld: Transcript, 139–162.

Barlösius, E. (2010). Ressortforschung. In: D. Simon, A. Knie, Andreas, S. Hornbostel (eds.) Handbuch Wissenschaftspolitik. Wiesbaden: VS, 377–389.

BGBl. Bundesgesetzblatt für die Republik Österreich.

Biegelbauer, P. (2007). Learning from Abroad: The Austrian Competence Centre Programme K plus. Science and Public Policy 34(9), 606–618.

Biegelbauer, P. (2013). Wie lernt die Politik? Lernen aus Erfahrung in Politik und Verwaltung. Wiesbaden: Springer VS.

BMWFW, BMVIT (2016). Austrian Research and Technology Report 2016. Wien: BMWFW/BMVIT.

BMWV (1997). K plus. Forschungskompetenz plus Wirtschaftskompetenz. Vorhabensbericht des Bundesministeriums für Wissenschaft und Verkehr zur Errichtung von Kompetenzzentren in Österreich. Wien: BMWV.

Braun, D. (1997). Die politische Steuerung der Wissenschaft. Ein Beitrag zum „kooperativen Staat". Frankfurt a. M./New York: Campus.

Braun, D. (2003). Lasting tensions in research policy-making – a delegation problem. Science and Public Policy 30(5), 309–321.

Braun, D. (2006). Delegation in the distributive policy area – the case of research policy. In: D. Braun, F. Gilardi (eds.) Delegation in Contemporary Democracies. London: Routledge, 146–170.

Braun, D. (2008a). Organising the political coordination of knowledge and innovation policies. Science and Public Policy 35(4), 227–239.

Braun, D. (2008b). Lessons on the political coordination of knowledge and innovation policies. Science and Public Policy 35(4), 289–298.

Braun, D., Guston, D.H. (2003). Principal-agent theory and research policy: an introduction. Science and Public Policy 30(5), 302–308.

Christensen, T., Lægreid, P. (2006). Agencification and Regulatory Reforms. In: T. Christensen, P. Lægreid (eds.) Autonomy and Regulation. Coping with Agencies in the Modern State. Cheltenham: Edward Elgar, 8–52.

de Boer H., Enders J., Schimank U. (2007). On the Way Towards New Public Management? The Governance of University Systems in England, the Netherlands, Austria, and Germany. D. Jansen (ed.) New Forms of Governance in Research

Organisations. Disciplinary Approaches, Interfaces and Integration. Dordrecht: Springer, 137–152.

Jochen G., Werner M. (1996). Anwendungsorientierung von Grundlagenforschung? Erfahrungen der Akademie der Wissenschaften der DDR. Frankfurt a. M./New York: Campus.

Groß, T., Arnold, N. (2007). Regelungsstrukturen der außeruniversitären Forschung. Organisation und Finanzierung der Forschungseinrichtungen in Deutschland, Baden-Baden: Nomos.

Groß, T., Karaalp, R.N., Wilden, A. (2010). Regelungsstrukturen der Forschungsförderung. Staatliche Projektfinanzierung mittels Peer-Review in Deutschland, Frankreich und der EU. Baden-Baden: Nomos.

Hohn, H.-W. (2010). Außeruniversitäre Forschungseinrichtungen. In: D. Simon, A. Knie, Andreas, S. Hornbostel (eds.) Handbuch Wissenschaftspolitik. Wiesbaden: VS, 457–477.

Hohn, H.-W., Schimank, U. (1990). Konflikte und Gleichgewichte im Forschungssystem. Akteurkonstellationen und Entwicklungspfade in der staatlich finanzierten außeruniversitären Forschung. Frankfurt a. M./New York: Campus.

König, T. (2017). The European Research Council. Cambridge: Polity.

Kritzinger, S., Pülzl. H. (2008). Governance Modes and Interests: Higher Education and Innovation Policy in Austria. Journal of Public Policy 28(3), 289–307.

Lepori, B., van den Besselaar, P., Dinges, M., Potí, B., Reale, E., Slipersaeter, S., Théves, J., van der Meulen, B. (2007). Comparing the evolution of national research policies: what patterns of change? Science and Public Policy 34(6), 372–388.

Orth, K. (2011). Autonomie und Planung der Forschung. Förderpolitische Strategien der Deutschen Forschungsgemeinschaft 1949–1968, Stuttgart: Steiner.

Pichler, R. (2014). Delegierte Koordination. Zur Koordinationsfunktion intermediärer Organisationen in der österreichischen Forschungsförderung. Österreichische Zeitschrift für Politikwissenschaft 43(4), 329–346.

Pichler, R. (2016). Agencification revisited. A case for the evaluation of FFG as an agency. fteval Journal for Research and Technology Policy Evaluation (41), 35–39.

Pichler, R., Hofer, R. (2014). Geschichte der kooperativen Forschung in Österreich. Die kooperativen Institute und ihre Dachorganisation zwischen Wissenschaft, Unternehmen und Politik. Innsbruck: Studienverlag.

Pichler, R., Hofer, R. (2016). Forschungskooperationen zwischen Hochschulen und Unternehmen im Wandel. Zur Entwicklung von Kooperationsformen, Rechtsrahmen und öffentlicher Förderung. In: W. Hauser (ed.) Hochschulrecht. Jahrbuch 2016, Wien/Graz: NWV, 407–418.

Pichler, R., Stampfer, M., Hofer, R. (2007). Forschung, Geld und Politik. Die staatliche Forschungsförderung in Österreich 1945–2005. Innsbruck: Studienverlag.

Pichler, R., Strimitzer, E. (2013). Ausgliederungen als Instrument eines modernen Verwaltungsmanagements: Grundlagen, Probleme, Praxis. In: E. Strimitzer, E. Thewanger, J. Bauer, M. Nayer (eds.) Handbuch des modernen Verwaltungsmanagements. Wien: Linde, 99–135.

Pollitt, C., Talbot, C., Caulfield, J. (2005). Agencies: How Governments Do Things Through Semi-Autonomous Organisations. Basingstoke: Palgrave Macmillan.

Rechnungshof (2004). Fonds zur Förderung der wissenschaftlichen Forschung, Forschungsförderungsfonds für die gewerbliche Wirtschaft. Reihe Bund 2004/4, Wien: Rechnungshof, 105–141.

Rechnungshof (2015). FFG und FWF – Interne Kontrollsysteme. Reihe Bund 2015/7, Wien: Rechnungshof, 89–199.

Skoie, H. (2000). Diversity and identity: the merger of five research councils in Norway, in: Science and Public Policy 27(2), 83–96.

Stampfer, M., Pichler, R., Hofer, R. (2010). The Making of Research Funding in Austria. Transition Politics and Institutional Development, 1945–2005. Science and Public Policy 37(10), 765–780.

Stucke, A. (1993). Institutionalisierung der Forschungspolitik. Entstehung, Entwicklung und Steuerungsprobleme des Bundesforschungsministeriums. Frankfurt a. M./New York: Campus.

Szöllösi-Janze, M. (1990). Geschichte der Arbeitsgemeinschaft der Großforschungseinrichtungen 1958–1980, Frankfurt a. M./New York: Campus.

Talbot, C. (2004). The Agency idea: sometimes old, sometimes new, sometimes borrowed, sometimes untrue. In: C. Pollitt, C. Talbot (eds.) Unbundled Government: A critical analysis of the global trend to agencies, quangos and contractualisation. New York: Routledge, 3–21.

Trischler, H., vom Bruch, R. (1999). Forschung für den Markt. Geschichte der Fraunhofer-Gesellschaft, München: C.H. Beck.

Van Thiel, S. (2012). Comparing Agencies across Countries. In: K. Verhoest, S. van Thiel, G. Bouckaert, P. Lægreid (eds.) Government Agencies: Practices and Lessons from 30 Countries. Basingstoke: Palgrave Macmillan, 18–26.

Wieland, T. (2009). Neue Technik auf alten Pfaden? Forschungs- und Technologiepolitik in der Bonner Republik. Eine Studie zur Pfadabhängigkeit des technischen Fortschritts. Bielefeld: Transcript.

Doris Schartinger, Bernhard Dachs

Innovation Co-operation between industrial firms and universities in Austria: Key concepts and developments over time

Abstract: *The scientific interest in science-industry relations has a long tradition. In this paper, we analyse one type of science-industry relation that enables learning-by-interacting to take place – innovation cooperation. Empirically the paper draws on the CIS 2014, which includes Austrian industrial firms that co-operate with diverse partners in Austria and in foreign countries.*

1 Introduction

Enterprises employ a variety of different inputs in their innovation processes. One of these inputs is knowledge acquired in co-operative projects from universities. Research in science-industry relations has a long tradition by now, with an increased attention during the last 25 years. The focal interest of these approaches always lies in the exploitation of academic insights which, in turn, yields financial rewards. This exploitation can happen in collaboration with industrial firms, who in their strive for competitiveness, interact with a variety of market and non-market organisations to facilitate their innovative performance. What is crucial for all interactions between these market and non-market organisations and institutions is the associated flow of knowledge.

Influential for the whole discourse on knowledge transfer had been the work by Nelson (1959), Arrow (1962), Dasgupta and David (1994) and Cowan, David and Foray (2000). Whereas the transfer and acquisition of some types of knowledge depend on prior codification or individual learning techniques, the transfer and acquisition of other types of knowledge necessarily imply inter-personal co-operation. This may be considered as a collective, interactive and iterative process between actors that build on heterogeneous individual, organisation-specific and region-specific resources. The heterogeneous knowledge backgrounds require face-to-face contacts in order for the processes of learning to take place. Although it may not be assumed in general that spatial proximity states a necessary condition in order for such knowledge spillovers to occur, as far as inter-personal co-operation is involved, spatial proximity seems to be advantageous in generating knowledge spillovers. Lengthy travels in order to establish face-to-face contacts may hamper knowledge diffusion in both quantity and quality due to high monetary and opportunity costs. This led to an extensive stream of literature discussing the geography of R&D collaborations (Scherngell 2013, Fischer and Varga 2003, Fischer and Getis 2010, Saxenian 2013, Saxenian 1990, Feldman, 1994).

Science-industry relations have received great attention in the empirical analysis of innovation systems at national, regional, sectoral and technological levels due to their focus on learning-by-interacting and the subsequent benefits of knowledge flows among diverse actors of innovation systems (Lundvall 1992, Nelson 1993, Edquist 1997, Malerba 2002, Furtado 1997, Fischer 2001). They have also been one of the hallmarks of the knowledge-based economy where "*the realisation that the diffusion of knowledge is just as crucial as its generation led to increased attention to knowledge distribution networks and systems of innovation*" (OECD 1996). Related to this, indicator-focussed discussions arose attempting to implement integrated sets of indicators around the different forms of knowledge (intellectual capital), the value-added processes and the results of the organisational knowledge-production processes, fundamental for the existence and flourishing of organisations. Relational capital, i.e. co-operation with different types of partner organisations, has always been a central element in the sets of indicators (Leitner 2004, Leitner and Warden 2004).

Today, science-industry relations are increasingly analysed under the lens of the open innovation paradigm. Here, open innovation is associated with a number of benefits where collaborations allow industrial firms to better adapt to dynamic market needs, share resources and distribute risks among partners, and, in the end, lead to higher commercial returns (Chesbrough 2003, Chesbrough et al. 2006, Du et al. 2014).

Furthermore, the concept of innovation systems in general and science-industry relations in particular have also benefitted from contributions from complexity theory (Fischer and Fröhlich 2001). Here, industrial firms and universities are seen as agents in a complex system which operate under high uncertainty, and engage in R&D networks in order to keep pace with innovation and disruptive change. This, in turn, creates interdependencies, i.e. relations and flows between the agents. These interdependencies foster dynamics and complexity in the respective (sectoral) systems (Vermeulen and Paier 2017), e.g. the biotech system (Korber et al. 2009).

This present study concentrates on innovation co-operation of Austrian firms. The empirical part is based on panel data derived from the various Community Innovation Surveys over the years.

In this paper, the following research questions will be addressed:

- How does innovation co-operation and science industry co-operation in particular evolve over time, and with respect to types of partners and geographical distribution of partners?
- Are there differences between innovation co-operation between Austrian industrial firms and universities compared to other types of partners?

The paper starts out with a general classification of different types of learning which are related to innovation co-operation. A detailed description of data sources for the empirical analysis is followed by the descriptive results concerning the spatial distribution of project partners on the level of countries and languages.

2 Learning in R&D co-operation

Modern economies are increasingly based on the creation, diffusion and application of knowledge and information. The success of many types of organisations and of economies as a whole in a globalised and complex economy is highly dependent on their capability to generate, gather, utilise and combine knowledge in an effective way. This renders learning and access to R&D networks a crucial ingredient to the innovation processes.

As there exist various types of knowledge, there also exist various mechanisms to transfer knowledge. The economic literature distinguishes between two types of knowledge: the codified and the tacit knowledge. Codification relates to the processes of reduction of knowledge and conversion into easily transferable pieces of information. The pre-existence of standards of reference, (numerical, symbolic, pictorial, geometrical languages, taxonomies of any kinds) and vocabulary of precisely defined and commonly understood terms facilitate the process of codification substantially (David and Foray 1995). It is this process of codification that causes the transmission, storage and reproduction of information to be less costly and makes it capable of being enjoyed jointly without diminishing, i.e. it renders non-rival. If the transmission and reproduction of a non-rival good is undesirable, costly measures must be taken to restrict access of those who want to use it without compensation or with compensation less than its value (Jaffe 1989, Jaffe 1996, Dasgupta and David 1994).

Although much knowledge is codified and publicly accessible, a major part of the essential knowledge is intangible. It has not been codified because it is organisation-specific, industry-specific or even individual-specific (*uncodified* knowledge) and has some degree of tacitness (Cowan et al. 2000). It cannot be easily transferred because it has not been stated in explicit form. One important kind of this implicit and partly tacit knowledge is human skills and techniques;

another kind has to do with the implicit but shared beliefs and modes of interpretation which make intelligent communication possible (Polanyi 1958). The acquisition of implicit knowledge is relatively costly in comparison to the transfer of codified knowledge and it may occur through various channels. On the one hand, implicit knowledge may be acquired through individual experientialism in the development of some capability or piece of knowledge by one single person. On the other hand, implicit knowledge may be acquired through the co-operation of at least two persons by activities such as talking and listening or demonstrating and copying (Machlup 1980).

A further distinction in the acquisition of implicit knowledge can be made between its development as a by-product of routine activities in an organisation, such as administration, production, distribution, consumption or marketing, and the conscious investment of time and resources in the expansion of the knowledge base of an organisation. Learning activities that are referred to in this context are, among others (see also Table 1):

- learning-by-doing (Arrow 1962) (increasing the efficiency of production operations),
- learning-by-using (Rosenberg 1982) (increasing the efficiency of complex systems),
- learning-by-imitating (Bandura 1976) (routinely learning from a model),
- searching (Lundvall 1988) (profit-oriented investigation of alternatives in terms of products, processes, markets, etc.),
- exploring (Lundvall 1988) (research with weaker goal-orientation than searching, sometimes results in outcomes that break cumulative paths and build the basis for new technological paradigms), and
- learning-by-interacting (Lundvall 1988) (some form of communication between individuals and organisations with heterogeneous knowledge backgrounds).

Table 1: The Acquisition of Implicit Knowledge – Classification of Types of Learning

	Routine Activities	**Knowledge Investment Activities**
Individual Experientialism	i) learning-by-doing ii) learning-by-using	iv) searching v) exploring
Inter-personal Co-operation	iii) learning-by-observing	vi) learning-by-interacting

The types of learning serve as examples, not taxatively.

Of course, these various types of learning mentioned here are enhanced by diverse preconditions (individual, organisational, competitive, etc.). And it may easily be derived that space does not state a matter of regard in all of these types of learning. In fact, it is only learning activities that involve a certain amount of inter-personal co-operation that can be assumed to be favoured by spatial proximity.

Knowledge that is transferred by inter-personal co-operation, i.e. by demonstration, personal instruction and the provision of expert services (advice, consultation, etc.) is heavily dependent on heterogeneity and face-to-face contact. If every individual had the same knowledge and background, no amount of communication would increase the stock of knowledge. However, individuals with different knowledge backgrounds from diverse disciplines or diverse experiences employ different cognitive schemata. In order to interpret and synthesise pieces of their differing knowledge, processes of questioning, trial, feedback and evaluation are likely to be necessary. Details as gestures, facial expression and tone of voice are crucial in the search for meaning and context. Through these social processes, eventually a shared language and common frames of reference are developed that promote learning and knowledge accumulation and facilitate innovation (Feldman 1994).

Therefore, it is heterogeneous and complementary knowledge that is essential in order for processes of learning to take place and that, in turn, requires face-to-face contacts to some extent in order to get these processes to work. Establishing face-to-face contacts over great distances imposes a considerable burden on every organisation in terms of monetary and opportunity costs. To the extent that face-to-face-contact is facilitated by geographical proximity through lower travel and opportunity costs, space should matter in types of learning processes that are conditioned by inter-personal co-operation.

It is the purpose of this paper to analyse one type of inter-personal co-operation that most often represents an investment activity by an organisation, and thereby enables learning-by-interacting to take place – this is the case of innovation co-operation. Units of observation are Austrian industrial firms that co-operate with diverse partners in Austria and in foreign countries. The Community Innovation Survey defines innovation co-operation as "*active participation with other enterprises or institutions on innovation activities. Both partners do not need to commercially benefit. This excludes pure contracting out of work with no active co-operation.*"

In this paper special attention is paid to university partners in innovation co-operation of Austrian industrial firms. In general, universities contribute to the production of knowledge by developing and providing new knowledge; to the transmission of knowledge by educating and developing human resources; and to the transfer of knowledge by disseminating knowledge and providing inputs to problem solving. The diffusion of knowledge from universities to other institutions may occur in a variety of ways. In addition to the development of human resources through education, knowledge flows are usually attributed to the use of faculties as technical consultants, post-graduate students as research assistants, the use of university facilities, informal communication between individuals, conferences, seminars, talks and social activities, or joint participation in commercial ventures by university and other scientists through contracted research projects (Acs et al. 1995). Furthermore, there is the possibility of knowledge flows through joint R&D projects, joint publications with members of other organisations, spin-off formations of new firms, scholarships and postgraduate linkages via the joint supervision of PhDs and master theses, exchange of personnel and sabbatical periods for university members, lectures at universities held by members of other organisations and training of employees by university members (Schartinger et al. 2001).

This analysis is focused on an aggregate level of knowledge transfer – innovation co-operation between industrial firms and universities and other types of partners – and the spatial distribution of these different types of partners. It is thereby assumed that innovation co-operation involves inter-personal relations between members of diverse organisations, which in turn enables learning-by-interacting. No reference can be made to what exactly is being learned, who learns it and how much is being learned. This problem is inherent in this particular area of research: The creation of knowledge and its diffusion are intangible activities which are difficult to define and measure. They are definitely not measurable in terms of what we normally think of as statistical variables (Grupp 1990). As an aggregate measure that is certainly based on person-to-person interaction it is related to the conceptualisation of academic engagement, established by Perkmann et al. (2013). According to them, academic engagement entails formal activities such as collaborative research, contract research and consulting, as well as informal activities like providing informal advice and networking with practitioners. It is seen as an important way of transferring academic knowledge into the industrial domain, significantly more valuable than licensing university patents.

Per definitionem innovation co-operation are no routine activities. In this context, innovation co-operation can be categorised as knowledge investment activities. In order to be of scientific interest for a university department, project partners should be equipped with complementary knowledge of any kind, which, in turn, requires some extent of face-to-face contacts between project partners in order to facilitate the interpretation and synthesis of differing pieces of knowledge.

3 Science-industry relations in Austria - empirical evidence

The following chapter will provide evidence on the development of science-industry relations in Austria over the long term and compare Austria with other European countries. The literature on science industry relations shows that there are huge differences in co-operation behaviour between different sectors and size classes within the business sector (Srholec 2011, Abramovsky et al. 2009, Schartinger et al. 2002, Dachs et al. 2008). We will examine these differences.

The section employs data from the Community Innovation Survey (CIS). CIS is a survey on the innovation activities of firms administered by Eurostat and national statistical offices. CIS results for Austria are available for eight different periods: 1994–1996 (CIS2), 1998–2000 (CIS3), 2002–2004 (CIS4), 2004–2006 (CIS2006), 2006–2008 (CIS2008), 2008–2010 (CIS2010), 2010–2012 (CIS 2012), and 2012–2014 (CIS 2014). The CIS sample includes mining, manufacturing, energy and water supply, wholesale, transport, information and communication, banking and insurance, and selected business services. The representative sample includes enterprises with 10 or more employees. Individual rounds of the CIS are based on a common definition of the main aggregates and are comparable over time. However, there are some differences in the sample composition, which may explain larger variations between CIS 2 and CIS 3, and between CIS 4 and later rounds (Falk and Leo 2004, 12).

Moreover, innovation has gained considerable prominence in public debates in Austria and at the European level in the 20 years. This fact has to be considered when we interpret the results over time, because it may constitute an upward bias over the years. There are also considerable differences in economic conditions between different CIS waves; average GDP growth was 3.5 % between 1998/2000, but only 1.1 % in the period 2002/04. During the financial crisis of 2008/2010, there was even a drop of 1.8 %, which is reflected in figures by a lower innovation propensity. Enterprises change their expectations of the future over the business cycle and, accordingly, adapt their innovation behaviour.

Co-operation became an important topic in Austrian science and innovation policy at the end of the 1990s. The overall share of co-operating enterprises reflects this fact only with a considerable delay (see Table 2). The share of co-operating firms remained quite stable for about 10 years between 1994/96 and 2002/04. We see a noticeable increase only in the period of 2004/06 and later.

Over the long term, there is an increase in the willingness to co-operate across all sectors and size classes. Manufacturing firms generally have a higher co-operation propensity than service firms, which is due to the fact that manufacturing firms innovate more often; if we only compare innovative firms, the differences between manufacturing and services disappear. In both groups, around 50 % of all innovative firms also co-operate.

Large firms co-operate more often compared to smaller firms, which is due to their larger capacities to maintain and monitor co-operation, and their higher absorptive capacities which are needed to absorb the knowledge provided by co-operation partners. The crisis of 2008/09 reduced the willingness to innovate and co-operate, but only with a considerable delay. 2010/12 is the period with the lowest co-operation propensity since 2006/08.

Table 2: Enterprises with innovation co-operation as a share of all enterprises

	1994/96	**1998/00**	**2002/04**	**2004/06**	**2006/08**	**2008/10**	**2010/12**	**2012/14**
Total	11.6 %	10.3 %	9.1 %	19.7 %	16.6 %	22.4 %	16.9 %	22.5 %
Manufacturing	13.1 %	9.9 %	10.8 %	20.2 %	18.4 %	25.5 %	19.5 %	26.7 %
Services	9.7 %	10.4 %	7.6 %	19.2 %	15.1 %	19.8 %	15.1 %	19.7 %
10–49 employees	8.6 %	7.4 %	6.6 %	15.2 %	11.8 %	16.9 %	12.6 %	16.7 %
50–249 employees	17.4 %	14.1 %	12.3 %	30.3 %	26.5 %	35.7 %	26.4 %	36.5 %
250 employees and more	34.6 %	34.7 %	40.6 %	55.1 %	54.7 %	61.3 %	50 %	62.3 %

Source: STATISTIK AUSTRIA. Only innovative firms.

We now turn to science-industry relations in Austria. The following table shows the share of co-operating firms in Austria according to type of co-operation partner. A firm can co-operate with different types of partners, so the shares do not sum up to 100 %. The data indicate that universities are among the most frequent co-operation partners of Austrian firms, only surpassed by suppliers and customers (which have been split into two categories in the two most recent surveys, so no single figure is available). The figures also indicate that the share of firms co-operating with universities is much more stable than other categories; the growth in the share of co-operating firms, therefore, is not due to more science-industry relations. However, in absolute terms there was an increase in co-operation with universities, because the share of innovative firms increased in the Austrian economy.

The data also allow to dig deeper and go to the sectoral level to see differences in science-industry relations between industries. Here, we see that manufacturing firms have a higher propensity to co-operate with universities than service firms. We find the highest share of university co-operations in medium and high-tech sectors such as chemicals and pharmaceuticals, electronics and automotive. But water supply and business services also have high shares of firms which co-operate with universities. The lowest shares of university co-operations are found in mining, textiles and clothing, in wholesale and in banking and insurance. Relevant knowledge in these sectors mainly comes from suppliers and customers.

Table 3: Share of co-operating firms according to type of co-operation partner

	1994/96	1998/00	2002/04	2004/06	2006/08	2008/10	2010/12	2012/14
Suppliers	38.8 %	54.8 %	43.0 %	58.6 %	56.6 %	53.5 %	58.9 %	56.2 %
Customers	40.5 %	43.7 %	44.8 %	59.8 %	41.9 %	43.3 %		
Competitors	22.6 %	48.5 %	22.5 %	36.0 %	23.8 %	21.5 %	34 %	29 %
Consultants, commercial labs			41.8 %	37.1 %	37.8 %	36.6 %	35.4 %	35.3 %
Universities	42.1 %	45.0 %	57.6 %	41.5 %	50.5 %	42.6 %	50.6 %	44.8 %
Other public research	21.5 %	20.7 %	30.1 %	23.8 %	18.7 %	18.1 %	31.1 %	23.6 %

Source: STATISTIK AUSTRIA. Only innovative firms; multiple counts allowed.

There are also important differences between small, medium-sized and large firms. Small firms (up to 49 employees) prefer to co-operate with suppliers and customers; universities are only the third most frequent partner type and have a similar share on all co-operation partners like competitors. Large firms with more than 250 employees, in contrast, have a clear preference for universities as co-operation partners. Three fourth of all large co-operating firms join with universities in innovation projects. The second most important partner type are suppliers.

How do Austrian enterprises compare to their peers in other countries in terms of co-operation? The data provided below indicate that the country does very well in international comparison: Austria is among the leading countries in Europe in terms of co-operation propensity, only surpassed by the UK, Estonia and Belgium, but well above EU average. Around half of all innovative Austrian enterprises co-operate, compared to a share of only 33 % at EU-level.

In terms of science industry relations, Austria performs even better with a share of 22.7 % of all innovative enterprises co-operating with universities. This is the second highest share in the EU, only surpassed by Finland. Which factors can explain this position? First, it appears from the table that co-operation and science industry co-operation in particular is high in many small countries; notable exceptions are only the UK and the Netherlands. So, the Austrian position may be explained by the size of the country and the limited number of actors in many fields, which makes it easy to identify the relevant players and connect with them. Moreover, the Austrian government made co-operation between science and industry a main priority of their funding activities, so there are considerable financial incentives to enter co-operation, which may not exist in other countries. These incentives may be the reason why Switzerland – which does not offer such incentives – rank only in the lower third of the table.

Austrian firms are very open to co-operation with domestic partners and within Europe. The excellent position of Austria, however, disappears once we look at co-operation with partners outside Europe. The last column of Table 4 reports the share of innovative enterprises in each country that co-operate with partners in India and China. In this indicator, Austria ranks only in the middle of the countries, while Sweden, Slovakia, and Denmark lead the table.

Table 4: Co-operations of Austrian enterprises in international comparison, 2012–14

Country	**Any type of co-operation**	**Co-operation with universities or other higher education institutions**	**Co-operation with a partners in China or India**
EU-28	33.1	13.2	:
United Kingdom	61.4	18.9	:
Estonia	57.0	14.6	5.0
Belgium	56.4	20.4	4.0
Austria	**50.6**	**22.7**	**2.9**
Slovakia	48.5	12.8	5.9
Lithuania	44.6	8.0	1.8
Slovenia	44.5	19.9	4.2
Norway	43.7	14.5	3.7
Greece	40.0	9.9	2.3

Country	**Any type of co-operation**	**Co-operation with universities or other higher education institutions**	**Co-operation with a partners in China or India**
Hungary	38.5	12.3	3.1
Netherlands	38.5	14.5	3.2
Denmark	38.2	15.4	5.6
Cyprus	38.2	6.0	4.1
Finland	38.2	23.0	5.0
France	35.8	12.2	2.8
Czech Republic	33.0	12.2	2.7
Sweden	32.9	15.3	6.9
Spain	32.1	10.9	1.4
Ireland	31.2	11.0	3.2
Romania	30.6	12.2	0.5
Croatia	28.7	8.0	2.1
Poland	28.2	10.6	1.7
Latvia	24.1	7.3	2.9
Luxembourg	23.9	11.0	3.5
Germany	21.8	14.1	1.3
Bulgaria	20.6	3.9	1.5
Italy	19.5	7.0	0.5
Portugal	19.2	9.2	0.9
Switzerland	17.9	9.9	1.5
Malta	15.4	4.0	2.2

Source: EUROSTAT. Multiple counts allowed.

An explanation may be the lack of Austrian multinational firms with operations in Asian countries, which may co-operate with Chinese and Indian partners in innovation projects. Austrian firms may also be less interested in co-operating with partners in these countries because they export to Asia mainly indirectly, for example as suppliers of the German automotive industry. Another possible explanation is concerned with the R&D and innovation activities of Austrian enterprises in China and India, which substitute co-operations between Austria and these countries. However, patent data do not indicate that Austrian firms perform a lot of these activities. So, the promotion of co-operation with non-European partners is a possible future priority of Austrian research, innovation and technology policy.

4 Conclusions

Industrial firms, in their strive for competitiveness, interact with a variety of market and non-market organisations to facilitate their innovative performance. Science-industry relations have received great attention over the years so that we can view them now from various theoretical and conceptual angles and with a variety of methodologies. Innovation, as well as science-industry relations are the topics that have gained considerable prominence in public debates in Austria and at the European level in the last two decades. This fact has to be considered when we interpret the results over time, because it may constitute an upward bias over the years.

Over the long term, there is an increase in the willingness of industrial firms to co-operate across all sectors and size classes. In terms of science-industry relations, Austria performs even better with over one fifth of all innovative enterprises co-operating with universities. This is the second highest share in the EU, only surpassed by Finland. Our data also indicate that universities are among the most frequent co-operation partners of Austrian firms, only surpassed by suppliers and customers. The figures also indicate that the share of firms co-operating with universities is much more stable than other categories; the growth in the share of co-operating firms, therefore, is not due to more science-industry relations.

Although national and even European links with external partners seem well established, Austrian firms are apparently more reluctant to engage in innovation co-operation with partners from countries outside Europe like China and India.

References

Abramovsky, L., Kremp, E., Lopez, A., Schmidt, T. & Simpson, H. (2009). Understanding Co-operative R&D Activity: Evidence From Four European Countries. *Economics of Innovation and New Technology,* 18, 243–265.

Acs, Z. J., Fitzroy, F. R. & Smith, I. (1995). Frontiers of Entrepreneurship Research http://www.babson.edu/entrep/fer/papers95/acs.htm.

Arrow, K. J. (1962). Economic Welfare and the Allocation of Resources for Invention (Reprint). *In:* STEPHAN, P. E. & AUDRETSCH, D. B. (eds.) *The Economics of Science and Innovation.* Cheltenham, UK. Massachussetts, USA.: Edward Elgar Publishing.

Bandura, A. (1976). *Lernen am Modell,* Stuttgart.

Chesbrough, H. W. (2003). *Open Innovation. The New Imperative for Creating and Profiting from Technology,* Boston, Harvard Business School Press.

Chesbrough, H. W., VanHaverbeke, W. & West, J. (2006). *Open Innovation. Researching a New Paradigm,* Oxford, Oxford University Press.

Cowan, R., David, P. & Foray, D. (2000). The explicit economics of knowledge codification and tacitness. *Industrial and Corporate Change,* 9, 211–253.

Dachs, B., Ebersberger, B. & Pyka, A. (2008). Why do Firms Co-operate for Innovation? A comparison of Austrian and Finnish CIS 3 results. *International Journal of Foresight and Innovation Policy,* 4, 200–229.

Dasgupta, P. & David, P. A. (1994). Toward a New Economics of Science. *Research Policy,* 23, 487–521.

David, P. A. & Foray, D. (1995). Accessing and expanding the science and technology knowledge base. *STI-Review,* 16, 13–68.

Du, J., Leten, B. & VanHaverbeke, W. (2014). Managing open innovation projects with science-based and market-based partners. *Research Policy,* 43, 828–840.

Edquist, C. (1997). *Systems of Innovation, Technologies, Institutions and Organisations,* Routledge.

Falk, M. & Leo, H. (2004). Die Innovationsaktivitäten der österreichischen Unternehmen. Empirische Analysen auf Basis der Europäischen Innovationserhebung 1996 und 2000. Vienna: WIFO.

Feldman, M. P. (1994). *The Geography of Innovation,* Boston, Kluwer Academic Publisher.

Fischer, M. M. (2001). Innovation, knowledge creation and systems of innovation. *Annals of Regional Science,* 35, 199–216.

Fischer, M.M. & Fröhlich, J. (2001). *Knowledge, Complexity and Innovation Systems,* Berlin, Heidelberg, Springer-Verlag.

Fischer, M.M. & Getis, A. (2010). *Handbook of applied spatial analysis: software tools, methods and applications,* Berlin Heidelberg, Springer-Verlag.

Fischer, M.M. & Varga, A. (2003). Spatial knowledge spillovers and university research: Evidence from Austria. *The Annals of Regional Science,* 37, 303–322.

Furtado, A. (1997). The French system of innovation in the oil industry – Some lessons about the role of public policies and sectoral patterns of technological change in innovation networking. *Research Policy,* 25, 1243–1259.

Grupp, H. (1990). Technometrics as a Missing Link in Science and Technology Indicators. *In:* SIGURDSON (ed.) *Measuring the dynamics of technological change.* London: Pinter Publishing.

Jaffe, A. B. (1989). Real effects of academic research. *American Economic Review,* 79, 957–970.

Jaffe, A. B. (1996). Economic analysis of research spillovers: Implications for the advanced technology program. http:/www.atp.nist.gov/www/eao/gcr708.htm.

Korber, M., Paier, M. & Fischer, M.M. (2009). An Agent-Based View of the Biotech Innovation System. *Cers 2009–3rd Central European Conference in Regional Science, International Conference Proceedings – Young Scientists Articles*, 1114–1132.

Leitner, K.-H. (2004). Intellectual capital reporting for universities: conceptual background and application for Austrian universities. *Research Evaluation,* 13, 129–140.

Leitner, K.-H. & Warden, C. (2004). Managing and reporting knowledge-based resources and processes in research organisations: specifics, lessons learned and perspectives. *Management Accounting Research,* 15, 33–51.

Lundvall, B.-A. (1992). *National Systems of Innovation,* London, Pinter Publishers.

Lundvall, B.-Å. (1988). Innovation as an Interactive Process: From User-Producer Interaction to the National System of Innovation. *In:* DOSI, G. (ed.) *Technical Change and Economic Theory.* London, New York: Pinter Publishers.

Machlup, F. (1980). *Knowledge: Its Creation, Distribution and Economic Significance,* Princeton, Princeton University.

Malerba, F. (2002). Sectoral systems of innovation and production. *Research Policy,* 31, 247–264.

Nelson, R. R. (1959). The Simple Economics of Basic Scientific Research (Reprint). *In:* STEPHAN, P. E. & AUDRETSCH, D. B. (eds.) *The Economics of Science and Innovation.* Cheltenham, UK. Massachussetts, USA.: Edward Elgar Publishing.

Nelson, R. R. (1993). *National Innovation Systems. A comparative analysis,* Oxford, Oxford University Press.

OECD (1996). The Knowledge Based Economy *OECD,* GD (96)102.

Perkmann, M., Tartari, V., McKelvey, M., Autio, E., Broström, A., D'Este, P., Fini, R., Geuna, A., Grimaldi, R., Hughes, A., Krabel, S., Kitson, M., Llerena, P., Lissoni, F., Salter, A. and Sobrero, M. (2013). Academic engagement and commercialisation: A review of the literature on university – industry relations. *Research Policy,* 42, 423–442.

Polanyi, M. (1958). *Personal Knowledge: Towards a Post-Critical Philosophy.,* London, Routledge & Kegan Paul.

Rosenberg, N. (1982). *Inside the Black Box. Technology and Economics,* Cambridge, Cambridge University Press.

Saxenian, A. (2013). The Origins and Dynamics of Production Networks in Silicon Valley. *Collaboration and Competition in Business Ecosystems,* 30, 283–309.

Saxenian, A. L. (1990). Regional Networks and the Resurgence of Silicon Valley. *California Management Review,* 33, 89–112.

Schartinger, D., Rammer, C., Fischer, M.M. & Fröhlich, J. (2002). Knowledge Interactions between Universities and Industry in Austria: Sectoral Patterns and Determinants. *Research Policy,* 31, 303–328.

Schartinger, D., Schibany, A. & Gassler, H. (2001). Evidence of Interactive Relations Between the Academic Sector and Industry. *Journal of Technology Transfer,* 26, 255–268.

Scherngell, T. (2013). *The Geography of Networks and R&D Collaborations,* Switzerland, Springer International Publishing.

Srholec, M. (2011). Understanding the heterogeneity of cooperation on innovation: Firm-level evidence from Europe. Oslo: TIK WORKING PAPERS on Innovation Studies No. 20111201.

Vermeulen, B. and Paier, M. (2017). Innovation Networks for Regional Development. Concepts, Case studies and Agent-Based Models, Springer International Publishing.

Bernhard Dachs, Georg Zahradnik

Global firms and national innovation systems: the case of Austria

Abstract: *We study the internationalisation of business R&D. The R&D expenditures of foreign-owned have increased in nearly all countries where data is available. Their share is highest in high-technology industries. The share of service industry is increasing fast. Austria is among the countries with the highest share of foreign-owned firms on business R&D expenditures.*

1 Introduction

Foreign-owned firms are among the top performers of research and development (R&D) in many countries. Thus, they have a considerable influence on the technological capabilities of countries, and, in turn, determine competitiveness to a considerable degree. This makes the internationalisation of business R&D a key dimension of science, innovation and technology policy. By internationalisation of R&D, we mean the fact that firms conduct research and development at locations outside their home countries. The internationalisation of business R&D is part of a larger trend towards internationalisation in science. Signs of this general trend are the rising shares of publications and patents involving persons from different countries; the increase in trans-border joint ventures; the mobility of researchers and students, but also increased policy efforts towards internationalisation at the transnational level ('Open Innovation, Open Science, Open to the World').

The internationalisation of business R&D is a relatively young phenomenon, although scientists, knowledge and artefacts have always crossed borders easily in economic history. The early literature tended to regard internationalisation of business R&D as an unlikely phenomenon, because of the strong linkages of large firms to universities and other research organisations in their home countries (Patel and Pavitt, 1991). Since the year 2000, a growing body of literature provides evidence that the internationalisation of business R&D is gaining momentum (UNCTAD 2005, Hatzichronoglou 2008, Shapira et al. 2009, OECD 2010, Hall 2010).

We measure business R&D internationalisation with data on the R&D expenditures of foreign-owned firms collected by national statistical offices in their R&D expenditures surveys. Iversen et al. (2016) describe this data source and the data collection in detail. The key variable of the analysis is inward BERD, which are the R&D expenditures of foreign-owned firms in a particular host country/ sector. Inward BERD and domestic BERD – the R&D expenditures by domestic firms – together constitute total BERD or the total business R&D expenditures in a particular country or sector.

The first section of this chapter presents the 'big picture' of R&D internationalisation – the global trends in the relationship between the main players at country level and at sectoral level. The section is based on Iversen et al. (2016). The following sections examine R&D internationalisation in Austria. We look at the evolution of inward BERD over time, diversity in terms of home countries of foreign-owned firms and R&D internationalisation in selected industries.

2 Global trends in business R&D internationalisation

We start with the 'big picture' of business R&D internationalisation. Figure 1 summarizes the relations between the EU, the US, Switzerland, Japan and China as measured by inward BERD. The countries included in the figure cover the lion's share of R&D expenditures of foreign-owned firms worldwide.

For the US, EU, Japan, China and Switzerland, inward BERD is split into the shares of US, EU, Japanese, Swiss firms and in the share of firms from the rest of the world. The size of each circle represents total inward BERD. Moreover, the figure illustrates the major relations between these countries and country groups with lines.

On the global scale, R&D internationalisation is dominated by the relationship between the US and the EU. Emerging economies, including China, play only a minor role. This is the biggest difference between the internationalisation of R&D and the internationalisation of production and the picture which emerges from the analysis of global value chains.

In 2013, US firms spent around 18 billion Euro on R&D in the European Union. R&D expenditures of EU firms in the US are about 22 billion Euro. Inward BERD from the US in the EU-28, and vice versa, accounts for more than half of all inward BERD worldwide, if inward BERD between EU member states is excluded. The US accounts for a lion's share of the R&D expenditures by foreign-owned firms world-wide.

Figure 1: Inward BERD between the EU, the US, Japan, China and Switzerland (2013, Mio Euro, current prices)

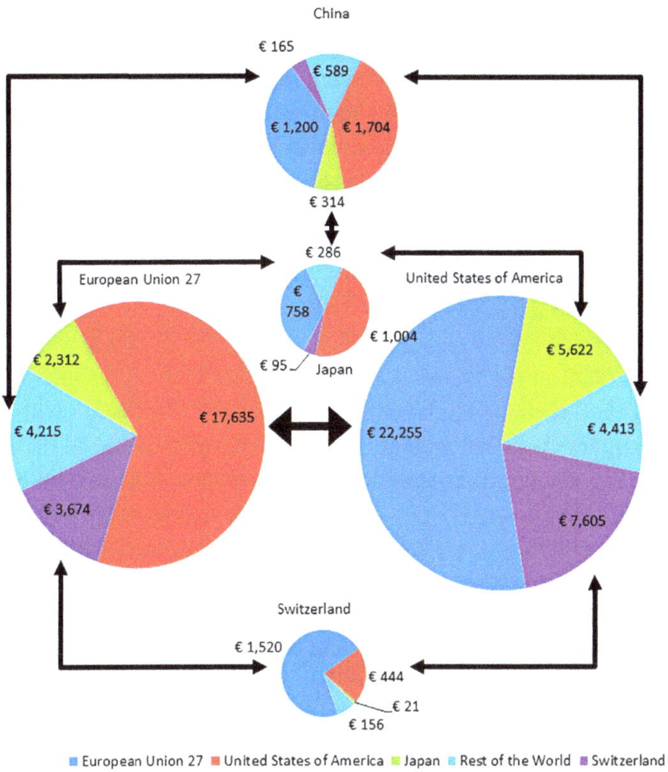

Notes: No country breakdown for China and Japan available, country shares based on PCT patent data (average 2011–2013)
Source: OECD, Eurostat, national statistical offices, own calculations

R&D expenditures of Japanese firms in the US (5.6 billion Euro) and the EU (2.3 billion Euro) as well as R&D expenditures of Swiss firms in the EU-28 (3.7 billion Euro), and in the US (4.5 billion Euro) are the most important remaining cross-border links created by inward BERD. Inward BERD from countries summarised under "Rest of the World" appears, if at all, almost entirely in the US or the EU-28. The single most important country included in the "Rest of the World" is Canada, accounting for about 0.9 billion Euro in the EU and another 0.4 billion in the US. Total inward BERD in China with about 4.3 billion Euro is significantly smaller than the corresponding inward BERD in the US or the EU.

However, there are major gaps in Chinese inward BERD data as there is no country breakdown available. PCT patent data on foreign-owned patents suggests that US R&D expenditures in China are about 40 % (1.7 billion Euro) and EU R&D expenditures in China are about 36 % (1.5 billion Euro) of total inward BERD in China. According to PCT patent data, Swiss and Japanese firms play only a limited role and likely have together less than 500 million Euro of R&D expenditures in China. Data on inward BERD in India is even less complete and not displayed in the figure. US outward data indicates higher growth rates for inward BERD in India compared to China.

Inward BERD from emerging economies including China and India is also included in "Rest of the World" for the EU and the US. While there are some gaps in the data, the magnitude of these investments is still very small. Chinese firms account for about 300 million Euro of inward BERD in the EU as well as in the US, while Indian firms have R&D expenditures of about 1 million Euro each in the EU and the US.

Compared with the picture in 2007 (Dachs, Stehrer and Zahradnik 2014), the shares of different countries and country groups on inward BERD at the global level are surprisingly stable. The big question, however, is the current and future role of China and India. There is no doubt that inward BERD in these countries has increased considerably and – in the case of India – is not completely captured in the data so far. Another open question is the role of R&D internationalisation in services, which is also increasing fast.

At the country level, Iversen et al. (2016) find that the internationalisation of R&D is rising slowly in the majority of countries where data is available. Inward BERD in absolute terms but also relative to total BERD increased in most countries. The share of foreign-owned firms is highest in small countries, and these countries are leading the process of R&D internationalisation. Inward BERD accounts for more than 50 % of total BERD in Slovakia, Ireland, Israel, Belgium, Romania, the Czech Republic, Greece, Hungary, the UK and Austria. Large countries such as Germany, Italy, France or the U.S., in contrast, show inward R&D intensities of only 17 % to 30 % of total BERD. Exceptions to this rule are some small countries, including Switzerland, Demark or Finland which show only a low level of R&D internationalisation. The United Kingdom, in contrast, is a large country with a very high R&D internationalisation.

Data provided in Iversen et al. (2016) also include information on the origin of inward BERD and the main home countries of foreign-owned firms. As a general trend, the share of the top investor country declines in the majority of countries between 2003 and 2013. Thus, the internationalisation of R&D became more diverse in the last decade and evolved from regional integration with neighbouring countries to more true international integration. Countries such as Romania, Hungary, Latvia, or Portugal are very much oriented towards the EU-28 and have virtually no inward BERD from non-European firms. The opposite is true for Malta, Ireland and Bulgaria. Between these two extremes, virtually every distribution between European and non-European firms can be observed. Belgium, France and Sweden lie in the middle of this distribution.

At the sectoral level, R&D internationalisation is concentrated in a few high-technology industries: chemicals; pharmaceuticals; machinery and equipment; computer, electronic and optical products; electrical machinery and apparatus; motor vehicles; and other transport equipment (including aircraft and spacecraft) account for the majority of inward BERD.

Pharmaceuticals is the largest sector (21.7 billion PPS Euro inward BERD in 2013 worldwide), and very much concentrated; the US plays a dominant role as a location for inward BERD in pharmaceuticals with a share of 65 %. Only two other countries, the UK and Belgium, account for more than 5 % of the total sectoral inward BERD.

The producers of computers, electronic and optical products attract a total of 9.5 billion PPS Euro inward BERD worldwide, ranking the sector as number three only behind pharmaceuticals and motor vehicles (11.4 billion PPS Euro). Again, the US is the single most important country, however, with a share of 35 % inward BERD, the cumulative share of the EU is well above this value with 58 %. Within the EU, more than 2/3 of inward BERD is concentrated in the three large countries: Germany (16 % of sectoral inward BERD worldwide), France (14 %) and the UK (10 %). The most important medium-sized country is in this sector is Austria with a share of 4 %. Electrical machinery and apparatus show a similar picture as computer, electronic and optical products in terms of the relative distribution across countries, but on a much smaller absolute level. Total sectoral BERD worldwide is only about 2.7 billion PPS Euro.

The motor vehicles sector is dominated by three large countries accounting for 3/4 of the sectoral inward BERD. With about 3 billion PPS Euro of inward BERD in motor vehicles in Germany, this is the largest absolute sectoral amount of inward BERD for any country except the US. As a result, Germany's share on total inward BERD is, with 26 %, only a little smaller than the US share of 31 %.

The third main country is the UK with a share of another 17 %; other notable countries are France (7 %), Spain, Czech Republic and Austria (each 3 %).

Data on the service sector is missing in some countries. From the available data, it appears that services are the second largest sector in terms of inward BERD and account for around a third of total R&D by foreign-owned firms. The US, UK and Canada are the most important host countries for service R&D internationalisation with a cumulative share of 70 %. Moreover, also some medium sized countries have significant shares, including Belgium (6 %) and Austria (5 %).

US data indicate that services are among the sectors with the fastest growth in R&D internationalisation. This expansion can be explained by various reasons. On the one hand, the use of new technologies makes service firms increasingly R&D intensive, like in the case of information and communication services. On the other hand, the R&D intensity in services increases because services and parts of the service value chain become increasingly tradable. As a result, manufacturing firms outsource R&D to specialized suppliers of R&D services. The most prominent example of this development is the pharmaceutical industry (Ramirez 2013) which moves clinical trials and other stages of the R&D process to specialized firms. Moreover, the emergence of small biotechnology firms has created a new type of divisions of labour between small and large firms in the pharmaceutical industry. Both developments lead to a higher internationalisation in services, and to more overseas R&D activities in service firms. Questions, however, remain about the service-specific drivers and obstacles of service firms, as well as the co-ordination costs and conditions for knowledge-transfer within service firms, which may be shaped by different degrees of tacitness compared to manufacturing.

To sum up, the cross-sectoral comparison of inward BERD reveals that inward BERD in various sectors is still concentrated in a small number of countries and the overall patterns is similar to the situation as of 2007 (Zahradnik 2014). However, we see a trend towards a wider variety of countries involved in the internationalisation of business R&D at the sectoral level. The sheer size of business R&D in the United States of America strongly influences the results.

3 The development of foreign-owned business R&D in Austria

In the second part of the chapter, we move our attention to business R&D internationalisation in Austria. Austria is a small, open European economy with strong economic ties to neighbouring countries. Added together, exports and imports equal Austria's GDP. Hence, in terms of internationalisation of trade, Austria can be compared to other small European countries such as Belgium, Finland, Sweden, Denmark, Ireland, the Czech Republic or Hungary (OECD 2010).

Austria is also highly internationalised in terms of terms of business R&D. Around half of the total business expenditures for R&D originate from foreign-owned firms (see Table below). This is one of the highest inward R&D intensities of all countries where data is available. This high share is mainly due to developments in the last decade. From 2003 to 2007, the R&D expenditures of foreign-owned firms (inward BERD) in Austria increased from about 1.2 billion Euro to 2.6 billion Euro, while domestic BERD only slightly increased from about 2.1 billion Euro to almost 2.3 billion Euro during the same time period. As a result, inward BERD rose in relative terms from 37 % of total BERD in 2003 to 52 % in 2007. The increase of total BERD in this period can therefore mainly be attributed to the activities of foreign-owned affiliates. However, it is important to note that according to representatives from Statistics Austria, this increase is at least partly caused by advances in identifying foreign-owned firms, particularly in the service sector.

Table 1: Inward BERD in Austria, 2003–2013

	2003	**2004**	**2007**	**2009**	**2011**	**2013**
Total BERD (Mio Euro)	3,309	3,556	4,846	5,093	5,693	6,778
Domestic BERD (Mio Euro)	2,095	1,959	2,261	2,447	2,844	3,345
Inward BERD (Mio Euro)	1,213	1,597	2,585	2,646	2,849	3,433
Inward BERD as % of total BERD	37 %	45 %	53 %	52 %	50 %	51 %

Source: Statistics Austria, own calculations

Internationalisation of R&D lost momentum after 2007 and inward BERD intensity decreased from 53 % to 50 % from 2007 to 2011. This was mainly due to a stronger growth of domestic BERD and may also be related to the financial crisis of 2008/09 which hit international economic activity severely. There was, however, no decrease in total inward BERD. The most recent data indicates that the growth of inward BERD has accelerated in the period 2011 to 2013. For the first time since the crisis, R&D expenditures of foreign-owned firms grew faster than R&D by domestic firms. This is a strong sign that Austria is still an attractive location for R&D of foreign multinationals.

4 Foreign-owned business R&D in Austria in a home country perspective

The distribution of inward BERD in Austria in terms of home countries of foreign-owned firms reveals two important facts. First, it shows that the internationalisation of R&D – despite the term – is still very much a regional phenomenon. Austria's neighbouring countries – most important, Germany and Switzerland – together account for 60 % of total inward BERD in Austria. This reflects the fact that distance between countries, cultural proximity, relative size of the home and host country, and the existence of foreign direct investment stocks between the two countries are decisive for R&D internationalisation (Leitner, Stehrer and Dachs 2013).

Table 2: Inward BERD from different home countries, Austria, 2003–2013

	2003		**2007**		**2009**		**2011**		**2013**	
	Mio Euro	**Share**	**Mio Euro**	**Share**	**Mio Euro**	**Share**	**Mio Euro**	**Share**	**Mio Euro**	**Share**
Germany	799	66 %	1,456	56 %	1,321	50 %	1,399	49 %	1,613	47 %
Netherlands	41	3 %	177	7 %	90	3 %	134	5 %	113	3 %
other EU	119	10 %	158	6 %	176	7 %	242	8 %	279	8 %
Switzerland	185	15 %	309	12 %	305	12 %	278	10 %	401	12 %
USA	30	2 %	228	9 %	421	16 %	385	14 %	529	15 %
Canada	16	1 %	180	7 %	192	7 %	234	8 %	298	9 %
other non-EU	22	2 %	76	3 %	141	5 %	177	6 %	200	6 %
Total	**1,211**	**100 %**	**2,585**	**100 %**	**2,646**	**100 %**	**2,849**	**100 %**	**3,433**	**100 %**
EU-27/28	**959**	**79 %**	**1,791**	**69 %**	**1,587**	**60 %**	**1,775**	**62 %**	**2,005**	**58 %**

Source: Statistics Austria, own calculations

Germany is by far the most important home country for inward BERD in Austria (see Table 2). German firms account for 1.6 billion Euro or around half of total inward BERD in Austria in 2013. This is double the amount of German inward BERD in 2003 at current prices. However, due to the large increase of R&D expenditures by firms from other countries, Germany's share on total inward BERD decreased from 66 % in 2003 to 47 % in 2013.

The second most important home country is the United States, with slightly more than 520 Mio Euro inward BERD in 2013. The corresponding share of the

US on total inward BERD is 15 %. US firms have overtaken Switzerland in terms of total inward BERD in 2009. Switzerland (401 Mio Euro) is now in third place. Another important non-European investor country is Canada (298 Mio Euro)

The focus on neighbouring countries decreases over time, in line with similar developments in other countries. The share of EU member states on total inward BERD decreased between 2003 and 2013 from 80 % to 58 %, while the share of non-EU countries increased proportionally. It should be noted that total inward BERD more than doubled between 2003 and 2013, so the decease of EU member states was in relative, but not in absolute terms.

The country group which grew fastest between 2003 and 2013 is 'other non-EU countries' excluding the US, Canada and Switzerland. It includes R&D expenditures by firms from China, (86 Mio Euro) the United Arab Emirates (49 Mio Euro), various offshore financial centres (36 Mio Euro) and a number of other non-European countries. These countries could more than triple their share, from 1.8 % in 2003 to 5.8 % in 2013. Total inward BERD for these countries amounts to 200 million Euro in 2013. No data is reported for India, South Korea, Turkey, or Israel on the country level, indicating no or very little R&D.

By comparing the shares of foreign-owned firms from different home countries on total inward BERD in Austria with the respective shares on total value added or employment (see Table 3), insights into the co-location of R&D and production can be gained. While the overall pattern is similar for R&D and production activities – Germany is of outstanding importance in both cases – R&D is much more concentrated compared to production. The top five home countries in terms of R&D account for 87 % of all foreign-owned business R&D, while the top five countries in terms of value added or employment only account for less than 75 %.

MNEs from a number of EU countries, including Italy, United Kingdom, Sweden, Spain and Denmark have significant production capacities in Austria but are of much less engaged in R&D. The same is the case for two non-EU countries, South Africa and Japan. At the same time, Swiss, Canadian and Chinese foreign-owned affiliates in Austria have much higher shares of total foreign-owned R&D compared to value added and employment. In all three cases, these large R&D activities are due to major investments in R&D intensive high- and medium-high-technology industries.

Neighbouring Czech Republic, Hungary, Slovakia and Slovenia have virtually no R&D in Austria and, with a combined share of total foreign-owned value added of under 0.5 %, also very limited production capacities in Austria.

Table 3: R&D expenditures, value added and persons employed by foreign-owned firms, Austria 2013

	R&D expenditure		Value added at factor cost		Persons employed	
	Mio EUR	Share	Mio EUR	Share	Number	Share
Germany	1,613	47.0 %	19,218	42.5 %	267,102	48.0 %
Netherlands	133	3.9 %	1,912	4.2 %	19,658	3.5 %
France	101	2.9 %	1,369	3.0 %	16,891	3.0 %
Italy	41	1.2 %	2,974	6.6 %	27,092	4.9 %
Luxembourg	39	1.1 %	528	1.2 %	6,945	1.2 %
United Kingdom	29	0.8 %	1,349	3.0 %	18,637	3.4 %
Sweden	15	0.4 %	1,010	2.2 %	11,448	2.1 %
Spain	5	0.1 %	423	0.9 %	7,535	1.4 %
Denmark	2	0.1 %	358	0.8 %	9,621	1.7 %
other EU-28	27	0.8 %	1,238	2.7 %	16,491	3.0 %
United States	529	15.4 %	5,697	12.6 %	52,484	9.4 %
Switzerland	401	11.7 %	4,050	8.9 %	50,418	9.1 %
Canada	298	8.7 %	1,413	3.1 %	14,446	2.6 %
South Africa	2	0.1 %	429	0.9 %	7,327	1.3 %
Japan	14	0.4 %	732	1.6 %	6,455	1.2 %
China	86	2.5 %	264	0.6 %	3,925	0.7 %
UAE	49	1.4 %	355	0.8 %	2,053	0.4 %
Offshore financial centres	36	1.0 %	807	1.8 %	9,773	1.8 %
other non-EU	13	0.4 %	1,136	2.5 %	7,978	1.4 %

Source: Statistics Austria, own calculations

5 Foreign-owned business R&D in Austria in a sectoral perspective

In terms of sectors, there was a remarkable shift in the structure of inward BERD in Austria between 2007 and 2013 from manufacturing towards services. A similar trend can also be observed at the international level. The first two columns of Table 4 show the shares of different sectors on total inward BERD in the two years.

Due to changes in the NACE classification of industries the shares of different sectors between 2007 and 2013 are not fully comparable. However, we can say for sure that, the share of the pharmaceutical and the electrical and electronics industry has decreased considerably, while the shares of communication services and commercial research and development services increased. In 2013, service industries account for about a third of total inward BERD, while the share of manufacturing industries dropped from 83 % to 68 %.

The growth of R&D of service firms is, on the one hand, triggered by the international expansion many service firms have started in recent years. On the other hand, it is also a sign of structural change; commercial R&D services also include biotechnology firms, so the relative decrease of pharmaceutical research by foreign-owned firms in Austria may reflect a shift from 'old' pharma to biotech. A similar shift may have occurred between electrical machinery and apparatus and the providers of information and communication services which also includes software services.

The last two columns of Table 4 indicate the share of inward BERD on total sector BERD. They show, in other words, which sectors of the Austrian economy are most internationalised in terms of R&D. We find the highest values in pharmaceuticals with a share of around 90 %, followed by motor vehicles. Nearly all other R&D intensive sectors have values of over 50 %, which indicates that the majority of R&D activities in these sectors are performed by foreign-owned firms. Machinery and equipment is the one large R&D performing sector with the highest share of domestic ownership, however, still half of all R&D in machinery and equipment is performed by foreign-owned firms.

The lowest inward BERD intensities are found in low-technology manufacturing sectors and in construction. Service sectors mostly reveal inward BERD intensities below or around 50 %, but are catching up fast. The comparison between 2007 and 2013 shows that inward R&D intensity has more than doubled in the largest knowledge intensive service sectors: information and communication and research and development services.

Table 4: Inward BERD share in different sectors, Austria, 2007 and 2013

	2007	2013	2007	2013
	Share on total inward BERD		**Inward BERD share on total sector BERD**	
Chemicals (excl. pharma)	4 %	3 %	71 %	61 %
Computer and related activities	4 %	12 %	36 %	54 %
Electrical machinery and apparatus	19 %	11 %	74 %	57 %
Information and Communication	n/a	7 %	23 %	53 %
Machinery and equipment	11 %	13 %	50 %	50 %
Medical and optical instruments	3 %	n/a	43 %	n/a
Motor vehicles	13 %	12 %	86 %	89 %
Other manufacturing	6 %	8 %	n/a	n/a
Other services	4 %	2 %	n/a	n/a
Other transport equipment	3 %	1 %	63 %	39 %
Pharmaceuticals	10 %	8 %	93 %	93 %
Radio, TV and communications	15 %	n/a	81 %	n/a
Research and development services	4 %	17 %	21 %	54 %
Wholesale, retail trade and repair	6 %	7 %	66 %	68 %
Total Manufacturing	**83 %**	**68 %**	**63 %**	**56 %**
Total Services	**17 %**	**32 %**	**31 %**	**44 %**

Source: Statistics Austria, Eurostat, OECD, own calculations

6 Summary

The internationalisation of R&D in the business sector is one of the forces that shape technological capabilities and the competitiveness of countries. This chapter reviewed the main trends of business R&D internationalisation.

On the global level, R&D internationalisation mainly takes place between the US and the EU – this is in sharp contrast to internationalisation via foreign direct investment or trade, where emerging economies including India and China have gained high shares in recent years.

R&D expenditures of foreign-owned firms have increased in absolute terms and as a share of total business R&D in nearly all countries where data is available. Small countries on average reveal a higher degree of business R&D internationalisation than large countries. At the sectoral level, R&D internationalisation mainly takes place in high-technology sectors. The share of services is increasing, leading to a higher degree of diversity at the sectoral level. We also see more diversity in terms of home and host countries: the share of the largest investor country on total inward BERD as well as the share of the largest host country of total outward is decreasing in the large majority of countries over time, leading the way to less dependence on one large (neighbouring) country and more true internationalisation.

The development in Austria resembles the trends at the international level. Austria is one of the most internationalised countries in the European Union in terms of inward BERD. Inward BERD amounts to 50 % of total BERD in 2013, which is one of the highest shares world-wide. Inward BERD in Austria mainly originates from two neighbouring countries – Germany and Switzerland – which together account for around 60 % of total inward BERD in Austria. The share of these two countries, however, is decreasing due considerable growth of R&D activities by non-EU firms, most notably the US. Asian countries have also a fair share in this development. Altogether, R&D internationalisation in Austria became less dependent on investment from other EU countries in recent years.

The sectors with the highest share of inward BERD are mostly high- and medium-high-technology industries – electrical machinery and apparatus, motor vehicles, computer, electronic products or pharmaceuticals. The share of service industries is growing fast in Austria.

References

Dachs, B., R. Stehrer and G. Zahradnik (eds., 2014). The Internationalisation of Business R&D. Cheltenham: Edward Elgar.

Eurostat (2012). Foreign Affiliates Statistics (FATS) Recommendations Manual, 2012 edition- URL: http://ec.europa.eu/eurostat/documents/3859598/5922981/KS-RA-12-016-EN.PDF/c93cdf48-5efa-459f.-b218-731a9a5476e9?version=1.0.

Hall, B. A. (2010). The Internationalisation of R&D, Berkeley: University of California. http://elsa.berkeley.edu/users/bhhall/papers/BHH10_RND_international_August.pdf

Hatzichronoglou, T. (2008). The Location of investment of multinationals linked to innovation, Paris: OECD.

Iversen, E., Dachs, B., Poti, B., Patel, P., Cerulli, G., Spallone, R., Zahradnik, G., Knell, M. and F. Lang (2016). Internationalisation of business investments in R&D and analysis of their economic impact (BERD Flows). Final report. Project for the European Commission, DG Research and Innovation.

OECD. (2010). Measuring Globalisation: OECD Economic Globalisation Indicators 2010. Paris: Organisation for Economic Co-operation and Development.

Patel, P., Pavitt, K., (1991). 'Large Firms in the Production of the World's Technology: An Important Case of "Non-Globalisation"'. Journal of International Business Studies 22, 1–22.

Ramirez, P. (2013). The new modes and geography of pharmaceutical R&D: Implications for policy, Birmingham: SERVICEGAP Discussion Paper 49.

Shapira, P., J. Edler, D. Gagliardi, A. Verbeek, E. Lykogianni, and M. Knell (2009). Analysis of R&D international funding flows and their impact on the research system in selected Member States, Brussels: RINDICATE specific assignment 7.

UNCTAD (2005). World Investment Report 2005: Transnational Corporations and the Internationalisation of R&D. New York and Geneva: United Nations. http://www.unctad.org/Templates/WebFlyer.asp?intItemID=2412&lang=1.

Zahradnik, G. (2014). R&D Internationalisation across Countries and over Time. In The Internationalisation of Business R&D, ed. B. Dachs, R. Stehrer, and G. Zahradnik.

Matthias Weber, Petra Schaper-Rinkel, Susanne Giesecke

Futures of research and innovation: transformative scenarios and the dilemma of research and innovation policy

Abstract: *The digitalisation of research and innovation activities, the enhanced participation of stakeholders and citizens in these, and the growing claims for social accountability of publicly funded research are posing major challenges for research, technology and innovation (RTI) policy. While expectations with regard to the contributions of RTI to help tackle major societal challenges are high, the complex nature of research and innovation constrains the possibilities of targeted policy intervention. Based on an overview of current trends in the practices and organisational forms of research, this chapter highlights some important tensions arising from the interactions between these trends, and sketches transformative scenarios that may be triggered by them. The scenarios point to fields of action that will require the attention of RTI policy in the years to come, as well as to new types of policy strategies to help shape the future.*

1 The twofold dilemma of research and innovation policy

Research findings and their implementation in the form of innovations are seen as having the potential to help tackle future challenges, be they economic (e.g. to promote growth), social (e.g. to fight inequality) or environmental (e.g. matters of security or climate change).

At the same time we are witnessing a profound process of change in the practices and organisational forms of research and innovation, which is reflected in such concepts as Science 2.0., Open Science, Big Data, Citizen Science, Responsible Research and Innovation (Szkuta and Osimo 2016, Leitner et al. 2016). Despite the contradictory nature of these developments (Schaper-Rinkel et al. 2013), it is to be expected that they will change our research and innovation systems and affect the possibilities and limitations of policy-making in this context.

In view of this, research and innovation policy in particular will play an important role, because it defines agendas and priorities as well as framework conditions and incentive mechanisms in order to shape, or participate in shaping future development paths. However, policy-makers face a twofold dilemma (Weber and Burgelman 2015): firstly, the fast pace of change in research and innovation calls for swift adaptation processes, while we know from experience that institutional changes need a lot of time to become effective. In other words: policy-making lags behind practice.

Secondly, there will be rising expectations of what research and innovation policy should achieve; the increased focus of state funding on research and innovation dealing with grand societal challenges may serve as proof. However, the complexity of the interaction of a larger number and variety of strategically active stakeholders at different levels makes targeted policy interventions more difficult and thus reduces the leverage of policy-makers. In other words: as expectations rise, the structural capacity of public policy to have an impact runs into its limits.

In what follows in this contribution, some of the current changes in research and innovation will be outlined and the resulting transformative potential will be explored by means of scenarios. For this purpose, we draw on results from the EU-funded project RIF Research and Innovation Futures 2030,[1] where the focus was more on changing research practices rather than innovation aspects. The transformative scenarios we will introduce are designed to shed light on the possibilities that arise from the new developments and, at the same time, highlight the sometimes significant differences from the current practice and organisation of research and innovation, with the transformation paths nonetheless being consistent and plausible.

It is likely that none of the scenarios taken individually provides a precise picture of the actual future of research and innovation, but rather that the scenarios taken together will reflect the range of possible futures. Nevertheless, when looked at together, they refer to a number of central fields of action that will be crucial for the future of research and innovation. They thus also provide an indication of which agendas should receive particular attention in future research and innovation policy.

2 Current trends in the practice and organisation of research

A number of developments and trends have been observed in a stocktaking exercise to identify fundamental changes in the practice and organisation of research activities, including changing forms of institutionalising research programmes. Six cross-cutting and interrelated dimensions have been identified (Schaper-Rinkel et al. 2013 and Weber et al. 2016):

1 The project "Research and Innovation Futures 2030. From Explorative to Transformative Scenarios (RIF)" was funded by the European Commission between 2011 and 2013 as part of the Science in Society Programme of the 7th Framework Programme for Research and Technological Development (project number SiS-2011–289058, www.rif2030.eu).

1. Digitalisation and virtualisation: information and communication technologies are changing research practice
   Digital technology makes codified information accessible worldwide and changes how scientists and other involved actors create, accumulate and disseminate knowledge. Science 2.0, also named e-science or cyber science, is characterized by the crucial role of interactivity and the future quality of data and information. The trend towards science 2.0 (Shneiderman 2008) includes changing practices in the lab, for instance the online availability of data through Open Notebook Science and it leads to new methods of developing hypothesis (data-driven research). Information and communication technologies are also profoundly changing academic publishing. Publications have integrated multimedia elements, online publications can be accessed more quickly and enable more rapid feedback mechanisms. Digital exchange and the reliable management of huge amounts of data call for special infrastructures and databases. This, in turn, creates many new challenges such as data access and security, quality assurance and intellectual property rights.
2. Collaboration and participation: new forms of collaborative research
   Collaborative forms of research have become standard in many disciplines and are supported by many different funding programmes. Research collaborations change because they are no longer implemented by the institutions but rather the researchers themselves, and Web 2.0 tools make communication, discussions and cooperation easy and immediate. This also opens up new opportunities for cooperation with citizens, by including patients in research projects for example, with citizens not just supplying the data, but being involved in the research (co-creation). More types of actors are now involved in research, which blurs the lines between science and society. What used to be described as interdisciplinary research has now been elevated to a new level by new technologies.
3. Emergence of new forms of access to data, infrastructures and research results
   Despite growing calls for openness, access to research results, data, infrastructures, grants and careers in research is still limited. There are different practices and they are guided by different visions of openness. Open Access and Open Science initiatives are increasingly supported by policy-makers (European Commission 2012). Many funding agencies on a European and national level have joined policy-makers in these demands. At the same time, there are a number of private companies (e.g. scientific publishers) that base their business models on commercial access to research data and analyses. The possibilities of more open access to research pose major challenges for traditional scientific

publishers in particular; this has led to various new funding models for the publication of research results.[2] However, in the longer term, entire new business models may develop independently of the publishers.

4. The changing societal impact of research
   In recent years emphasis has been placed on the demonstrable benefits of research and its social impact. Most recently the main focus has been on the contribution of research on tackling societal challenges. This can enhance the standing of interdisciplinary and transdisciplinary research, which is also reflected in the relevant funding opportunities. The external demands placed on research, however, contrast with evaluation systems within research that are still based on traditional and predominantly disciplinary systems. Bibliometric indicators are increasingly used to assess the quality of research, but they are also continuously adapted and improved in order to address the many points of criticism levelled at such standardised evaluation practices.
5. The globalisation and internationalisation of research
   Research and science have been increasingly carried out on an international level for some time. This is both due to global research agendas (climate research, for example) and new technological possibilities that facilitate collaboration. International cooperation agreements contribute to this development, as does global access to scientific literature (The Royal Society 2011). This development has been accompanied by an enormous increase in the mobility of researchers; in Europe in particular, the opening of the national research systems played an important role in this respect. This is further underlined by the significant increase in research spending by multinational corporations outside their home countries.
6. Re-contextualizing of science in society
   The relationship between science and society is changing (Owen, Macnaghten et al. 2012). Science is expected to play a key role in society's response to emerging global grand challenges over the coming decades (ICSU 2011; The Royal Society 2011; Keenan, Cutler et al. 2012). At the same time, as the boundaries between science and society become more open and increasingly blurred, the forms of scientific autonomy of science and public accountability are in the process of renegotiation. Despite the high expectations regarding

2 The two main models are the "gold" and the "green" standard. The gold standard is based on the publication of scientific articles in open access media and journals; the fees are paid by the researcher, his/her own organisation or the funding agency. The green standard grants open access through archiving preprints and reprints in a community-based web infrastructure (such as arXiv).

citizen science as potentially democratizing science and policy (cf. Ottinger 2010) citizens are often mainly used as a resource that is broadly available and cheap. Companies in the area of genetic testing for example, try to involve their customers in research activities by using the concepts and promises of citizen science. The most prominent example is the field of "citizen science genomics" (Swan, Hathaway et al. 2010). Finally, the question of what research and innovation should do for society is reconsidered; as a result, stakeholders become more involved in the development of (public) research and innovation programmes. This development goes hand in hand with a growing rationalisation and planning of science, and includes the continuous monitoring and assessment of research activities to ensure their social benefit and to justify the use of public funds.

One of the major results to come out of the analysis of developments and trends within the STI-system has been the enormous heterogeneity in the ways in which scientific research is done and organised. The issues of change we identified show contradictions that may lead to significant tensions. These potential contradictions and tensions are the starting point to anticipate different futures of research in the future.

3 From tensions to transformative scenarios

Scenarios are an effective means to capture alternative ideas of the future within a certain range; however, in order to develop transformative and, at first glance, seemingly unlikely visions and transformation paths for the future, special methodological approaches have to be employed that integrate creative and systematic elements. Furthermore, a wide range of different viewpoints must be taken into consideration to leave the beaten track and to be able to explore the different potentials. This can be done through workshops with present and future stakeholders of research and innovation.

In the RIF project, a two-stage process was implemented,[3] the first stage of which used emerging trends in order to develop so-called exploratory scenarios on a continuation of current trends. The aim of this first stage was to anticipate tensions that could be triggered by the trends. This approach is based on the argument that developments such as those outlined in the previous section are often in conflict with the prevailing historically grown institutional settings and organisational structures, or that incompatibilities can arise between these trends.

3 Cf. Erdmann et al. (2016) for details on how to develop transformative scenarios.

This argument is especially relevant when it is about such rapid developments as currently observed in research and innovation. The necessary institutional and organisational adjustments are often only made once deficits and massive conflicts that can no longer be dismissed make a fundamental process of change inevitable.

The identified tensions thus serve as a starting point for developing transformative scenarios in a second stage. The task here is to develop a plausible argument for the widespread implementation of a radical change process. Visible results that are achieved through alternative research practices and thus demonstrate the advantages over established practices can trigger the employment of new models and organisational forms by others. Through further imitation and adaptation, the predominant organisational models and practices can be questioned and the relevant transformative changes can be realised.

By way of example, the tensions that were used as a starting point for the development of transformative scenarios in the RIF project will be outlined below. They are of course only part of the wide range of possible tensions:

- *Tension 1: Fragmentation vs. the rising need for coordination:* While the increased focus of research and innovation on global or social changes requires a high degree of coordination between the stakeholders in research, at the same time an increasing fragmentation of research fields and conflicting actor strategies can be observed. This contradiction has a number of consequences, for instance with regard to the stakeholders' willingness to cooperate in the raising of research funds, or the definition of long-term research agendas.
- *Tension 2: Established science vs. alternative approaches:* Further tension emerges between established science and competing approaches to knowledge generation, which are gaining increasing importance in many areas. Examples include the many types of alternative medicine and the agrarian knowledge of indigenous peoples. In extreme cases, pseudo-scientific approaches may assume greater importance. This means there is tension not only when it comes to the different epistemic communities in science, but also about whether decisions should be based solely on 'modern' scientific findings or whether they can also be based on other forms of knowledge. These different forms of knowledge, or rather their representatives, fight for the legitimacy and credibility of and funding for their respective approaches.
- *Tension 3: Increasing social demands on research vs. the limitations of established research and innovation funding:* Significant tension can arise as a result of growing discontent in society about the accomplishments of established research and innovation policy (or research and innovation in general) if it does not manage to live up to its mission of tackling major societal challenges. As a

consequence, major long-term scientific projects of both basic and problem-oriented research could increasingly be questioned and thus also their ability to achieve fundamental breakthroughs in research.

- *Tension 4: Increasing economic pressure on the research system vs. openness in research:* Another dilemma could arise from the increasing economic pressure exerted on the research system. Fiercer competition for scarce funding limits willingness to cooperate and to openly exchange knowledge. Regular evaluations force researchers to publish research results as quickly as possible to ensure they maintain their personal profile, while cooperative research agendas are more difficult to put into practice in a competitive climate.
- *Tension 5: Rising demand for research vs. decreasing attractiveness of a career in science:* Increasing pressure to perform, an unsatisfactory work-life balance in academia, and uncertain prospects mean that careers in science are becoming less attractive. While there continues to be a great need for research, its image has changed considerably. Research is now less marked by depth of thought, creativity and a high level of freedom, but more by standardised tasks and strict administrative control.

These briefly outlined areas of tension form the starting point for the transformative scenarios shown in Table 1. They refer to a time period after 2030, as it is assumed that the transformation processes will take some time. While none of the scenarios can claim to be a complete picture of the future of R&I, they provide important alternative models that may exist side by side in different segments, or may be more relevant to some areas of research than to others. The scenarios "Open Research Platforms" and "Grand Challenges for Real" for example, show different ways of approaching societal challenges through research and innovation. At the same time, the scenario "Knowledge Value Chains" is an industry focused model for research and innovation that can also aim at tackling grand societal challenges. The scenario "Researchers' Choice" concentrates on quality issues in research and on motivations from the perspective of the individual researcher. It thus provides important bottom-up elements of future research and innovation, which complement key features of the scenario "Open Research Platforms". Finally, the scenario "Knowledge Parliaments" focuses on an alternative way of developing and making decisions about research agendas with the participation of actors and stakeholders, also as a complement to "Open Research Platforms". This shows that the road ahead could consist of an intelligent combination of elements from these five scenarios. They thus serve as inspiration for the design of a forward-looking strategy that points beyond established paths of thinking and the continuation of current trends.

Table 1: Transformative scenarios 2030

"Open Research Platforms" – Self-governance in a decentralized research landscape	The continuing fragmentation and ever more urgent need for coordination require new models to organise research for tackling societal challenges (tension 1). The first scenario, "Open Research Platforms", describes a research and innovation future of self-governance in a networked and decentralised research environment. Given the successes of the first global and open research platforms in medicine, this model was soon taken up in other fields. By the year 2030, the research landscape with its research-performing organisations (and individuals) and funding mechanisms will be largely decentralised, global and open. Research is initiated by virtual communities which are integrated into publicly accessible virtual platforms. The self-governance of research via Open Research Platforms (ORPs), which are accessible to universities, RTOs, industry, individuals, foundations and civil society organisations, will be the norm. The knowledge flows that pass through these ORPs are embedded in the soft coordination efforts of governments and open societies worldwide. These may relate to the monitoring of research, the support of research activities and specific incentives for researchers, in order to strengthen certain ORPs that are in the public interest.
"Knowledge Parliaments" – free negotiation of knowledge claims worldwide	Dwindling resources and the ongoing crisis of the traditional science system lead to fierce debates about the distribution of the resources between various established and alternative forms of knowledge and the communities that support them (tensions 2 & 3). New forums are created to negotiate the positions and claims regarding knowledge in society. Civil society organisations take on a stronger role in the development of public R&I programmes. By 2030, the various actors will negotiate all types of knowledge-related debates in so-called knowledge parliaments. These knowledge parliaments will become a new model for the governance of science in society. They prioritise research topics and provide forums for their negotiation, where actors with specific research interests, specific research topics and epistemologies compete for acceptance. This type of exchange also enables the formation of consortia on specific research topics. Citizens and many other local stakeholders and epistemic cultures (such as laypeople and local knowledge) can be integrated, and neglected research topics and unconventional knowledge domains can be brought to the fore. This will give rise to a global research environment that accepts the plurality of knowledge as well as the associated research practices or regulations concerning intellectual property.

"Grand Challenges for Real" – collective experimentation in socio-technical laboratories	While the major societal challenges are the focus of public R&I policy in Europe, on a global level China and other players are in the lead. The established research funding models in Europe have not delivered on their promises; new approaches are needed (tension 3). This scenario describes a future that arises around new transdisciplinary research practices within the context of collective experiments in socio-technical laboratories which are able to show quick results. In the year 2030, the research environment in Europe will be marked by the large-scale implementation of collective experiments. Research on the grand challenges in Europe will be organised around twelve major knowledge and innovation communities; each of them will manage various regional and local socio-technical laboratories where many different solutions to the grand challenges will be developed and tested. These socio-technical laboratories are key elements for facilitating learning processes to realise effective solutions. Various actors – citizens, companies, universities and social entrepreneurs – engage in such collective experiments. Experimentation, measurement of practices and impacts, and processes of co-creation go hand in hand, thus achieving remarkable progress in tackling the grand challenges.
"Knowledge Value Chains" – integrated research for innovation	The global race for leadership in technological innovation leads to intensified competition for funds for application-oriented and commercially relevant research (tension 4). The focus is on speed, efficiency and the reliable planning of major research initiatives. In this scenario, research and innovation are closely linked. By 2030, the public research landscape will be closely linked with the global private research landscape. In different European countries and regions, research will progress at different speeds with the aim of improving competitiveness in global markets through innovation. Research will be carried out in Knowledge Value Chains (KVCs), which coordinate collaboration between three types of highly specialised and complex organisations: research-integrating organisations (RIOs), which control the KVCs; research service organisations (RSOs); and specialised research suppliers (SRS). New planning methods are used to rationalise research and integrate it directly into industrial innovation agendas. National governments will strengthen their RSOs in order to position them as key players in the KVCs and as direct partners in the small number of RIOs.

"Researchers' Choice" – autonomous researchers strive for creativity and well-being	Increasingly precarious conditions at universities and research institutions and their decreasing reputation means that young scientists increasingly look for employment opportunities outside the established science system (tension 5). They are the independent and well connected drivers in this scenario. Creativity, autonomy and a reasonable quality of life are central values and replace the pursuing of traditional career paths. Society is marked by pronounced individualistic values and a strong focus on personal wellbeing, autonomy and creativity. Autonomous researchers are at the heart of scientific research. They will offer their scientific expertise and knowledge on virtual research markets where new business models emerge. To put their ideas into practice, researchers can choose between a wide range of options, which include anything from new forms of science-based entrepreneurship to more collective forms (summarised under the term "slow science") with a strong focus on local societal needs. In addition to global research initiatives, autonomous researchers can also get involved in local communities to tackle challenges and to gain recognition for their work.

4 Fields of action for future research and innovation policy

The scenarios open up a range of options that initially has to be considered from a normative perspective. The assessment of the various scenarios will differ depending on the actor and suggest different strategies for pursuing a future considered desirable and avoiding less desirable ones. This will result in a cluster of strategy elements for the actors which allow them to prepare for the future possibilities and play a role in shaping them.[4]

There are political agendas such as the establishment of the *European Research Area*. Given the openness of future development paths and the diversity of distributed actor strategies, however, it is almost impossible to ascertain which policy approaches would lead to the desired result. From a policy perspective, only three types of actions can be identified: i) where a general consensus can be found between the relevant actors, ii) where there is a need to mediate conflicts of goals and interests, or iii) where public policy intervention can leverage potential synergies between the actor strategies.

Specifically, the approach within the RIF project allowed the identification of six central fields of action for (European) research and innovation policy which

4 See Amanatidou et al. (2016), where the assessments and strategies of the various actor types are spelled out in more detail.

should be given increased attention in the coming years for the three reasons mentioned:

1. Science and society
   Several of the scenarios outlined imply a significant change in the role of science and research in society. They are expected to make a greater contribution to tackling societal challenges at a local, national or international level. This gives societal actors a greater say in the definition of agendas, the allocation of resources and the implementation of research activities. Knowledge is increasingly produced jointly by scientists and other stakeholders by working on complex tasks in an iterative process. To this end, new forms of cooperation and dialogue between science and society have to be tested so that research is both of high quality and high relevance to important societal issues. New criteria, therefore, need to be developed and implemented to legitimise and evaluate publicly funded research.
2. Research careers 2.0
   In a variety of ways, the scenarios indicate that in the future, research will no longer take place in established organisations only, but will penetrate society as a whole. This has an enormous impact on the career models for researchers and on the expertise and competencies they will get. Careers within the traditional science system may become the exception and, given the difficult conditions, may no longer be regarded as desirable. It is conceivable that virtual research communities and new systems of incentives and funding for young talents will become more attractive than the established organisations. This is not to say that there will no longer be universities and research institutions, but they are likely to significantly open up to the outside world. The pathways will be quite different, depending on whether the main driver of change will come from multinational companies or societal organisations.
3. A distributed and diversified funding landscape
   The diversification of funding opportunities for research raises issues not only about the stability and sustainability of research agendas, but also about the tensions and (in)-balance between different types of research (e.g. curiosity-driven vs. challenge-driven) and about the involvement of the various stakeholders in the definition of the relevant research agendas. New options will need to be tested regarding the transparency and legitimation of public funding for research.
4. Intellectual property rights and open access
   The various forms of openness reflected in the scenarios are already under intense discussion today. There are distinct and justified conflicts of interest,

because private interests are in conflict with the idea of knowledge as a public good. More differentiated approaches may be needed to do justice to the different constellations of interests in various research domains, while also utilising the benefits of open access. A current example for the future relevance of openness is the field of ubiquitous robotics and automation. Open knowledge, open access and open infrastructures as the basis of innovation in automatization and robots can become a crucial factor: European infrastructures such as public platforms and digital commons might become drivers for open innovation and might reduce the dependency of innovators from multinational corporations that limit their individual innovation capacity as well as the share in the profits and in tax revenue in Europe.

5. Governance of knowledge infrastructures
   Access to knowledge infrastructures plays an important role in all scenarios; however, similar to intellectual property rights and open access, the increasingly heterogeneous nature of research calls for a differentiated approach. Ethical, legal, social and security aspects will increasingly have to be taken into consideration. And finally, this also includes the broader issue of quality assurance, which will need to be reassessed in view of the new possibilities afforded by the social networks and increasingly heterogeneous research practices.
6. Policy and global research agendas
   Finally, the scenarios also imply a stronger integration of the European research landscape in new global structures of collaboration, which increasingly takes place outside research and innovation policy (e.g. through virtual platforms). In the context of research on global challenges in particular, this raises questions about the international integration of (European) research and the potential contribution of research and innovation policy. An open question is the rationality behind global research agendas and how to articulate global societal needs in research agendas beyond technology push or market pull.

5 Possibilities and limitations of research and innovation policy for the year 2030

The starting point of this contribution was the observation that research and innovation policy faces a twofold dilemma. First, there is a fast pace of changes in research and innovation and a comparative inertness of institutional structures; secondly, there are rising expectations on policy and limited means to intervene in the ever more complex world of research and innovation. The scenarios provide a number of basic suggestions on how policy-makers can deal with this twofold dilemma.

We have argued that emphasising individual trends is not enough to gain a differentiated picture of the future of research and innovation. Trends only create certain new opportunities that can, but do not have to, develop in interaction with other developments and in the context of systemic tensions and conflicts. The inertia of existing structures and institutions can get in the unfolding of these possibilities (and in some cases, this is perhaps just as well). The structural and institutional room for manoeuvre is often only expanded once their shortcomings can no longer be ignored in view of the emerging deficits and conflicts.

One of the key findings was that the range of plausible futures of research and innovation is extremely wide. There is a great number of qualitatively different scenarios that are already inherent to an extent in the current trends and can evolve from the tensions that were considered.

This diversity gives space to normative considerations, both from the perspective of the individual stakeholders and from a collective point of view. However, these normative considerations do not give rise to simple guidelines for policymakers. The traditional policy-making tools can no longer be used to control the increasingly diverse and decentralised strategies of research and innovation actors. Decisions are, to a large extent, determined by factors and developments that lie outside the state-controlled institutional and funding framework.

Nonetheless, some insights can be gleaned for research and innovation policy. First of all, greater attention should be placed on unexpected and at first glance seemingly radical developments, in order to develop approaches at an early stage to prepare for these new developments. In view of the diversity and further openness of future developments in research and innovation, opportunities have to be created that allow us to experiment with new research and innovation practices. At the same time, the resilience of research organisations must be strengthened; this will increase their ability to absorb and integrate new models of research and create a balance between adaptability and continuity. New and potentially transformative changes must be continuously explored and assessed to be able to quickly respond to new developments. Taken together, these indications point to a need for a quite fundamental and structural rethinking of how research and innovation policy is conceived, and how it is connected to other policy areas (van Oost et al. 2016).

Finally, the temporality of public policy intervention is important. This applies in particular to transformative developments which can unfold and overcome established structures against the background of significant tension and in combination with successful alternative practices. Public policy intervention can have a great impact and give direction to the transformation process during exactly this turbulent stage. A crucial question is, which normative frameworks can support this new direction and can guide future pathways for establishing new practices, organisational models and institutional framework conditions in coordination with the relevant research and innovation actors. A promising approach for a broad vision of inclusive science, technology and innovation is already codified in the Sustainable Development Goals (SDGs) of the United Nations, to which EU members and most other nations in the world committed themselves in a 2015 UN resolution. The EU has expressed support for the vision as an important driver of smart, sustainable and inclusive growth in the post-2015 framework.

References

Amanatidou, E., Loveridge, D., Saritas, O. (2016). "Strategies for emerging research and innovation futures", Foresight 18(3): 253–275.

Erdmann, L., Schirrmeister, E. (2016). "Constructing transformative scenarios for research and innovation futures", Foresight 18(3): 238–252.

European Commission (2012). Commission Recommendation on Access to and Preservation of Scientific Information, C(2012) 4890 final, European Commission, Brussels.

ICSU (2011). ICSU Foresight Analysis Report 1: International science in 2031 – exploratory scenarios. Paris, International Council for Science.

Keenan, M., P. Cutler et al. (2012). "Orienting international science cooperation to meet global 'grand challenges'". Science and Public Policy 39(2): 166–177.

Leitner K.-H., Warnke, P., Rhomberg, W. (2016). "New forms of innovation: critical issues for future pathways", Foresight 18(3): 224–237.

Ottinger, G. (2010). "Buckets of Resistance: Standards and the Effectiveness of Citizen Science". Science Technology & Human Values 35(2): 244–270.

Owen, R., P. Macnaghten et al. (2012). "Responsible research and innovation: From science in society to science for society, with society." Science and Public Policy 39(6): 751–760.

Schaper-Rinkel, P. et al. (2013). Exploring the future of research. Trends and drivers in doing and governing research, RIF Research Report, Vienna, http://www.rif2030.eu/wp-content/uploads/2013/02/2012_11_21_RIF-D1-1-Stocktaking_Report_korrigiert_BR.pdf.

Shneiderman, B. (2008). "Computer science – Science 2.0". Science 319(5868): 1349–1350.

Swan, M., K. Hathaway et al. (2010). "Citizen science genomics as a model for crowdsourced preventive medicine research". J Participat Med 2: e20.

Szkuta, K., Osimo, D.(2016). "Rebooting science? Implications of science 2.0 main trends for scientific method and research institutions", Foresight 18(3): 204–223.

The Royal Society (2011). Knowledge, networks and nations: Global scientific collaboration in the 21st century London, The Royal Society.

Van Oost, E., Kuhlmann, S., Ordonez-Matamoros, G., Stegmaier, P. (2016). "Futures of science with and for society: towards transformative policy orientations", Foresight 18(3): 276–296.

Weber, K.M. (2009). FTI-Politik im Spiegel von Theorie und Praxis: von Planung über Steuerung zu Governance. In: Leitner K.-H., Weber M., Fröhlich J. (eds.): Innovationsforschung und Technologiepolitik in Österreich: Neue Perspektiven und Gestaltungsmöglichkeiten, Studienverlag, Vienna, pp. 231–254.

Weber, K.M., Burgelman, J.-C. (2015). Die Zukunft von Forschung und Innovation. In: Rat für Forschung und Technologieentwicklung (ed.): Die Gestaltung der Zukunft. Wirtschaftliche, gesellschaftliche und politische Dimensionen von Innovation, Echomedia Buchverlag, pp. 550–584.

Weber, K.M., Amanatidou, E., Erdmann, L., Nieminen, M. (2016). "Research and innovation futures: exploring new ways of doing and organizing knowledge creation", Foresight 18(3): 193–203.

Tools and methods for the analysis of complex systems

Thomas Scherngell, Laurent Bergé and Iris Wanzenböck

Bridging centrality: A new indicator to measure the positioning of actors in complex networks[1]

Abstract: *In the recent past, we can observe a growing interest in the notion of positioning indicators, shifting emphasis to actors in the innovation process and their R&D inter-linkages with other actors. In relation to this, we suggest a new approach for assessing the positioning of the actors coming from the research on complex networks. We rely on the notion of bridging centrality (BC), based on the concept of bridging paths, i.e. a set of two links connecting three actors across three different aggregate nodes (e.g. organisations, regions or countries). The main argument is that triangulation in complex networks is a key issue for knowledge recombinations and the extension of an actor's knowledge base. As bridges are most often not empirically observable at the individual level of research teams, we propose an approximated BC measure that provides a flexible framework for dealing with the aggregation problem in positioning actors. Hereby, BC is viewed as a function of an aggregate node's (i) participation intensity in the network, (ii) openness to other nodes (i.e. the relative outward orientation of network links), and (iii) diversification of links to other nodes. In doing so, we provide an integrative perspective that enables us to achieve a better understanding of the positioning of certain actors in R&D networks. An illustrative example of the co-patent network of European regions demonstrates the performance and usefulness of our BC measure for networks constructed at the aggregated level, i.e. regions in our example. A region's outward orientation and the diversification of its network links moderate the influence of regional scale on network centrality. This is a major strength of the measure, and it paves the way for future studies to examine the role of certain aggregate nodes, and, by this, contributes to the debate on positioning indicators in the STI context*[2].

1 Introduction

Over the past decade, we have observed considerable progress in the advancement and application of Science, Technology and Innovation (STI) indicators (see, e.g., OECD 2005). In this context, the notion of *positioning indicators* has come into fairly wide use in the STI community. It originates from considerations on

1 This work has been supported by the Austrian National Bank (ÖNB), Jubiläumsfonds, project No. 16301.

2 This study is in large aspects based on Scherngell et al. (2016) and Bergé et al. (2015).

new requirements imposed to the production of STI indicators in terms of their adaptation from classical input-output to a positioning indicators framework, focusing on flows and linkages between research actors in the innovation system (Lepori 2008). These linkages materialize in form of more formalised collaborations in R&D such as joint R&D projects (see, e.g., Scherngell and Barber 2009 and 2011, Scherngell and Lata 2013), joint publication activities (see, e.g. Glänzel and Schubert 2004), and researchers mobility (see, e.g. Edler et al. 2011). Similarly, they may appear as informal knowledge flows – often referred to as disembodied knowledge spillovers (see, e.g., Fischer et al. 2006). Despite the fact that a number of works have provided evidence on the crucial importance of R&D linkages, most STI indicators remain rooted in a classical, linear conception of the innovation process.

In this study, we employ a complex networks perspective on R&D linkages. A *complex network* features – in contrast to smaller, non-complex networks – usually a very large number of nodes and links with non-trivial topological features. In our focus on R&D, respective networks under consideration comprise a set of nodes representing knowledge producing actors inter-linked via edges representing knowledge flows. In Social Network Analysis (SNA), the *positioning* of actors is usually described by the concept of *centrality*, trying to capture a certain function and/or role a node, *i.*e. actor, takes by means of its inter-linking to other nodes (Borgatti 2005). Thus, the concept of *centrality* can be very well related to the notion of *positioning indicators*. Up to now, only a few STI studies have started to utilize the centrality concept to capture the positioning of actors in innovation systems, usually employing most basic analytical concepts, such as degree centrality or betweenness centrality (see, e.g., Wanzenböck et al. 2014 and 2015, Heller-Schuh et al. 2011). However, they somehow neglect conceptual problems that arise if networks are defined at the aggregate level of (large) organisations or even regions and countries, rather than the most relevant level in terms of R&D linkages, usually a research team within an organisation. Furthermore, the used centrality measures are not conceptually adapted to the STI context, such as the incorporation of theoretical considerations on the diversification of links or the relative outward orientation of R&D networks.

Thus, we propose a measurement approach that specifically adapts common centrality measures to STI relevant theoretical considerations, and provides a flexible framework to overcome problems related to node aggregation of the network. We shift attention to the notion of *Bridging Centrality* (BC), based on the concept of inter-nodal *bridging paths*, denoting an indirect connection between two nodes via a third 'bridging node'. However, since bridges are usually difficult to be observed at the level of individual researchers, our proposed BC measure shows properties that allow us to estimate the centrality of an aggregate node based on the underlying micro-structure of the network. The objective of this study is to introduce the conception and formal derivation of BC, and demonstrate its interpretative power by an illustrative example. Our approximation of the number of bridging paths of an aggregate node is given by the product of three components which are, (i) a node's *participation intensity* in the network, *i.e.* its number of links, (ii) a node's *relative outward orientation*, *i.e.* the ratio of node-internal (loops) vs. node external links, and (iii) the *diversification* of links across other nodes in the network. Emanating from our conceptual discussion, we provide a formal proof on how our measure, which is decomposed into these three components, converges mathematically to a node's expected number of bridges. Since all three components are relevant to STI studies on its own, the measure shows high interpretive power and, by this, significantly enriches our toolset of positioning indicators, not only in terms of a more appropriate centrality measure.

The remainder of this study is structured as follows: The next section introduces the network notation to elaborate in some more detail on the notion of bridges. Afterwards, we outline the formal definition of our BC measure based on the three components: *participation intensity*, *relative outward orientation* and *diversification*. It shows how we conceptually perceive the number of bridges of a node starting from these three components, before we provide a formal proof that our measurement approach mathematically corresponds to the expected number of bridges of a node. Then, we shift attention to an illustrative example, where we apply our measure to the European co-patent network observed at the level of NUTS-2 regions, and compare the results with conventional centrality measures as well as with respect to the three BC components. The final section closes with some concluding remarks and ideas for a future research agenda.

1.1 Network definition and the concept of bridges

In social sciences, analytical strategies employed to deal with the divide between individualistic and holistic approaches for describing social systems are referred to as multilevel analysis (see, e.g., Lazega and Snijders 2015). In traditional sociological literature, this is aptly described as the phenomenon of *ecological fallacy*, pointing to logical failures in the inference of statistical data observed at an aggregated level on the nature and characteristics of individuals (see Robinson 1950). Social Network Analysis (SNA) on the one hand faces similar problems when applied to aggregate nodes, in particular in a STI context (see, e.g., Wanzenböck et al. 2014), while on the other hand entails promising potentials to overcome such analytical problems (Snijders 2016).

We argue that these aggregation problems prominently occur in the measurement of the positioning of actors in STI studies. Shifting attention to a network analytic context, we draw on the rich SNA toolbox to evaluate the positioning of actors. The concept of centrality is fundamental in this respect, usually adopted to assign a value to each actor depending on their position within the network (Wasserman and Faust 1994). However, most SNA measures of centrality have been developed for the analysis of social systems, where the nodes of the network are usually identified in terms of individual persons. Accordingly, the original meaning borne by the SNA centrality measures, as well as respective interpretations, relies on assumptions on the social behaviours of individual persons, and how these persons might influence each other by these behaviours. Centrality measures based on observations at the aggregate level (e.g. organisations, regions or countries), therefore, raise important conceptual issues. Most importantly, it implies that every individual actor of an *aggregate node* would homogeneously benefit from the R&D linkages to other nodes, irrespective of who establishes the relations and the strength of these relations.

We propose a flexible analytical approach to address conceptual problems related to unobserved micro-level structures of the observed network. Core in this context is the concept of '*bridging path*' denoting a form of indirect connection between aggregate nodes. For a formal definition, consider a network observed at the level of aggregate nodes, e.g. organisations, regions or countries, and the connections between the aggregate nodes represent the R&D linkages between their individual actors. This represents a weighted network where we define g_{ij} as the number of R&D linkages (*i.e.* micro-level links) between aggregate nodes i and j. Further, each micro-level link between two aggregate nodes is denoted by y_{ij}^a, representing the ath link between aggregate nodes i and j with $a \in \{1, \ldots, g_{ij}\}$. A pair of links (y_{ik}^a, y_{jk}^b) forms a bridging path if, and only if, y_{ik}^a and y_{jk}^b are connected to the same actor of aggregate node k. This concept is depicted by Figure 1 exemplified with three aggregated nodes.

The concept of bridges is of particular relevance in a STI context. A high number of bridging paths imply a more open positioning in the network. In contrast to closed and dense network structures, such a bridging position between other nodes can be related to the access to a more diversified knowledge pool. It is assumed that the sources from which the individual actors draw their knowledge will have an impact on their ability to generate innovations, and knowledge flowing through bridging paths is more likely heterogeneous and non-redundant.

Based on these conceptual considerations on bridging paths, we propose a measurement approach for *Bridging Centrality (BC)* in the section that follows.

Figure 1: Illustration of the notion of bridging path

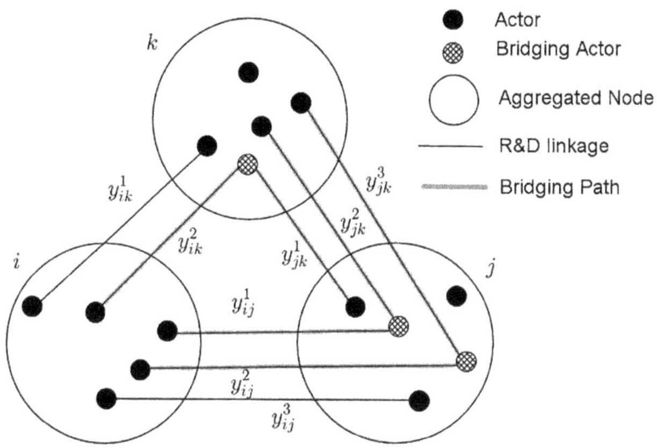

Notes: The figure depicts three bridging paths formed by the following pairs of links: (y_{ik}^2, y_{jk}^1), (y_{ij}^1, y_{jk}^2) and (y_{ij}^2, y_{jk}^3). So the aggregate node dyads (j, k), (i, k) and (i, j) have respectively 0, 2 and 1 bridging paths stemming from aggregate nodes i , j and k.

2 Three components of Bridging Centrality

Given the parsimonious and effective formal definition of bridging paths, it could be assumed at a first glance that the definition of a formal BC measure is straightforward. Indeed, this is the case in pure mathematical terms as the *true* measure of BC for an aggregate node i would just be the number of bridging actors assigned to i, probably normalised by the total of all bridging actors in the network. However, we are de facto often confronted in social sciences with a well-known problem of finding appropriate empirical observations for the objects under scrutiny. This is particularly critical in STI studies, where we usually focus on large-scale networks such as co-patent, co-publication or project networks. Most often information on the links at the level of the individual researchers cannot be traced; even when information is available (as e.g. for authors in publications and/or inventors in patents), the observation for large-scale networks is infeasible due to immense efforts for data cleaning, in particular name standardisation over space and time.

Thus, we propose an alternative measure for BC that approximates the number of bridges of an aggregate node. Drawing on theoretical considerations from various STI studies, we assume that the number of bridging actors of an aggregate node may, to a large extent, depend on three components: the node's (i) *participation intensity,* (ii) *relative outward orientation* and (iii) *diversification* of network links. We will show that a linear-multiplicative combination of these components formally dissolves to the *expected number of bridges* (see Bergé 2016). At the same time, the three components of BC are highly relevant; each of them has important mechanisms on its own and significant implications on knowledge creation structures.

In our formal description, we denote C_i as the approximated BC for the aggregate i node by

$$C_i = q_i \, s_i \, (1 - h_i) \quad (1)$$

where

q_i is the weighted degree of aggregate node i, defined as the total number of links excluding node-internal ones, i.e. $q_i = g_i - g_{ii}$. It refers to the overall *participation intensity* in the network; an aggregate node's size will amplify the probability of yielding more bridges between other nodes.

s_i is the *relative outward orientation* of aggregate node i with $s_i = q_i / g_i$. It reflects the degree of openness of an aggregate node with respect to all established links. Given the focus on bridges, the capacity of an aggregate node to link to other nodes would decrease by a higher number of node-internal links (loops) as it potentially reduces the number of actors connecting different aggregate nodes.

h_i refers to the degree of *diversification* of network links of aggregate node i among other nodes. It is measured by the Herfindahl-Hischman (HH) index by $h_i = \sum_{j \neq i} (g_{ij} / q_i)^2$. The term $1 - h_i$ varies between 0 and 1, and indicates how an aggregate node's linkages are distributed along its neighbouring nodes in the network. The more the linkages are concentrated, the less the node is central in terms of BC. Concentration reduces the actors' possibility to build bridges among different aggregate nodes and to draw its knowledge from different sources.

An aggregate node's ability to benefit from new ties in the R&D network or exploit external knowledge sources via the links may be determined by all three components together. Outward orientation and higher diversification in particular may help actors belonging to an aggregate node to develop and renew their knowledge base faster, or prevent lock-in situations in certain technologies (see, e.g., Breschi et al. 2015). Hence, our measure to approximate BC features promising opportunities in terms of interpretation.

However, it is not only conceptually attractive, but also mathematically corresponds very well with the *Expected Number of Bridges* (ENB) measure as introduced by Bergé (2016) using basic random matching assumptions between aggregate nodes[3]. Mathematically, our measure simply collapses to ENB, given by

$$\begin{aligned} C_i &= q_i\, s_i\, (1-h_i) = \frac{q_i^2}{g_i}\left(1-\sum_{j\neq i}(g_{ij}/q_i)^2\right) = \frac{q_i^2}{g_i} - \frac{1}{g_i}\sum_{j\neq i} g_{ij}^{\;2} \\ &= \frac{1}{g_i}\sum_{j\neq i} g_{ij}(q_i - g_{ij}) = \frac{1}{g_i}\sum_{j\neq i}(g_{ij}\sum_{k\neq i,j} g_{ik}) \\ &= \sum_{j\neq i}\sum_{k\neq i,j}\frac{g_{ij}g_{ik}}{g_i} \end{aligned} \qquad (2)$$

with the latter expression corresponding to the most general form of the expected number of bridges. This can of course be extended to more reasonable assumptions of expected bridges, for instance, by considering the number of actors of aggregate nodes proportional to the number of R&D linkages (see Bergé 2016 for details).

3 An application to European cross-region patent networks

In this section, we shift attention to a compact illustration of the BC measure in order to demonstrate its behaviour and interpretative power. We use the example of co-patent network data that is observed at the level of European NUTS-2 regions to represent the aggregate level in our network, and compare the BC with three other commonly used centrality measures, that is the degree, the eigenvector and the betweenness centrality (Wasserman and Faust 1994)[4]. A co-patent is regarded

---

3 Note in this context that the random matching process is 'noisier' the larger the aggregate node is (e.g. countries).

4 The degree is here calculated as the number of unique R&D interactions the agents of a region are involved in. The eigenvector and the betweenness centrality are computed using the package igraph available in the statistical software R. Both these two measures are based on the weighted regional co-patent network where the nodes are

as a collaboration of at least two inventors issuing a patent grant, providing us a trail of R&D linkages. Respective data are extracted from the REGPAT database (see Maraut et al. 2008) and consist of all patents applied for at the European patent office (EPO) in the period 2006–2010. Our cross-regional co-patenting network is based on a total of 171,451 patents, producing 121,036 inter-regional collaborations linking 250 NUTS-2 regions (see Bergé et al. 2015 for further details on the data)[5]. In this example, the aggregation problem described in Section 2 clearly applies as we are not able to observe co-patent activities at the level of individual inventors[6].

Table 1 represents the top 30 centralities ordered by the bridging centrality. We focus on commenting the most salient differences. The ranking is clearly dominated by German regions which rank highest for most measures[7]. However, the concentration tendency and high clustering of co-patenting activities at the national level of Germany may point to the fact that economic linkages at the national level prevail. Likely explanations are low cultural barriers as well as lower transaction costs. These factors seem to promote the high regional bridging centrality in German regions[8].

---

the regions and where the linkages between any two regions are the number of patents co-invented by agents from these two regions. Due to the nature of the network, we used the weighted version of both the betweenness and the eigenvector centrality.

5 Note that the use of different time frames to build the dataset, such as 2004–2006 or 2008–2010, imply no important changes on the results.

6 An aggregation to the organisational level would also be inconsistent, as patents are most often assigned to headquarters of companies, which often does not reflect to the locus of knowledge creation.

7 Note that the performance of German regions is not merely driven by the fact that German NUTS2 regions are usually smaller geographical aggregates than NUTS2 regions in other EU countries, which could drive up their number of inter-regional collaborations at the national level. Indeed, when we redo the analysis taking German regions at the NUTS1 level while keeping other regions at the NUTS2 level, German regions still trust the top of the rankings.

8 The national versus international nature of collaborations and its effects on regional network centrality might deserve further attention, and constitute an interesting route for the further development of the regional bridging centrality measure.

Table 1: Centrality values of the top 30 regions for the co-patent network (ranks in brackets)

NUTS-2 Code	Region	BC (C_i)	Degree Centrality	Eigenvector Centrality	Betweenness Centrality
DE12	Karlsruhe	1.00 (1)	0.87 (5)	1.00 (1)	0.22 (10)
DE71	Darmstadt	0.93 (2)	0.88 (4)	0.82 (3)	0.45 (4)
DEA1	Düsseldorf	0.84 (3)	0.82 (6)	0.68 (4)	0.22 (9)
DEA2	Köln	0.76 (4)	0.73 (7)	0.63 (6)	0.33 (6)
DEB3	Rheinhessen-Pfalz	0.73 (5)	0.64 (8)	0.85 (2)	0.13 (16)
DE21	Oberbayern	0.63 (6)	0.96 (2)	0.42 (7)	1.00 (1)
DE11	Stuttgart	0.59 (7)	0.95 (3)	0.64 (5)	0.37 (5)
DE13	Freiburg	0.49 (8)	0.52 (10)	0.34 (9)	0.19 (11)
CH03	NW Switzerland	0.43 (9)	0.41 (14)	0.16 (17)	0.10 (24)
DEA5	Arnsberg	0.42 (10)	0.39 (16)	0.33 (10)	0.06 (45)
DE14	Tübingen	0.40 (11)	0.44 (12)	0.38 (8)	0.07 (36)
DE30	Berlin	0.39 (12)	0.40 (15)	0.22 (14)	0.19 (12)
DEA3	Münster	0.39 (13)	0.31 (20)	0.27 (11)	0.05 (49)
DE25	Mittelfranken	0.37 (14)	0.43 (13)	0.20 (15)	0.11 (20)
CH04	Zurich	0.35 (15)	0.34 (18)	0.12 (21)	0.08 (32)
FR10	Île de France	0.34 (16)	1.00 (1)	0.08 (35)	0.93 (2)
DE27	Schwaben	0.33 (17)	0.31 (21)	0.25 (12)	0.03 (71)
DE40	Brandenburg	0.28 (18)	0.22 (30)	0.15 (18)	0.05 (54)
DE60	Hamburg	0.27 (19)	0.23 (29)	0.09 (28)	0.05 (48)
DE26	Unterfranken	0.27 (20)	0.27 (23)	0.25 (13)	0.10 (23)
FR42	Alsace	0.26 (21)	0.23 (27)	0.13 (19)	0.09 (31)
CH02	Espace Mittelland	0.26 (22)	0.27 (22)	0.08 (30)	0.05 (50)
BE24	Vlaams-Brabant	0.25 (23)	0.20 (34)	0.04 (46)	0.10 (25)
DE92	Hannover	0.24 (24)	0.25 (24)	0.12 (22)	0.05 (53)
FR71	Rhône-Alpes	0.24 (25)	0.57 (9)	0.08 (34)	0.33 (7)
DEB1	Koblenz	0.21 (26)	0.17 (46)	0.18 (16)	0.01 (96)
DE93	Lüneburg	0.21 (27)	0.17 (42)	0.07 (37)	0.02 (79)
CH05	Eastern Switzerland	0.21 (28)	0.19 (36)	0.07 (38)	0.01 (97)
BE21	Prov. Antwerpen	0.20 (29)	0.18 (38)	0.05 (44)	0.09 (28)
BE10	Région de Bruxelles	0.20 (30)	0.14 (59)	0.03 (55)	0.08 (34)

Further interesting specific cases are, e.g., Île de France (FR10) or Brussels (BE10). FR10 ranks at the 16th position for BC, while being ranked first with respect to its degree centrality. Degree centrality may overstate its position; Although the structure of the collaborations of FR10 with its partnering regions is highly distributed (it has a low HH index of 0.04), this region is characterised by a high number of internal collaborations (the outer share of collaborations is only 44 %), and thus, does not provide many bridging paths to the inter-regional R&D network. BE10 ranks below 55th for degree and eigenvector centrality, while for BC it ranks 30th. These conventional centrality measures may underestimate its positioning in the inter-regional co-patent network due to its very high outward orientation (its outer share is 94 %) and a highly distributed structure of collaborations (it has a low HH index of 0.07); BE10 is likely to provide many bridging paths to the network.

To complement the results from Table 1 in this respect, Table 2 provides a snapshot on top-5 regions including Île de France (FR10) and their respective results across the three components. It becomes obvious that the high rankings of German regions result from the fact that they show both a high participation intensity and openness, i.e. a high absolute as well as relative number of inter-regional co-patents.

Table 2: Ranking of top regions decomposed by three components of BC

Rank	NUTS2	BC	Participation Intensity	Outward Orientation	Diversification
1	DE12	1.00	2333	0.85	0.88
2	DE71	0.85	2050	0.78	0.93
3	DEB3	0.80	1831	0.92	0.84
4	DEA1	0.80	1993	0.80	0.87
5	DEA2	0.78	1866	0.84	0.87
13	FR10	0.34	1382	0.49	0.96

Figure 2: Bridging centrality of regions in the European co-patent network decomposed by three components

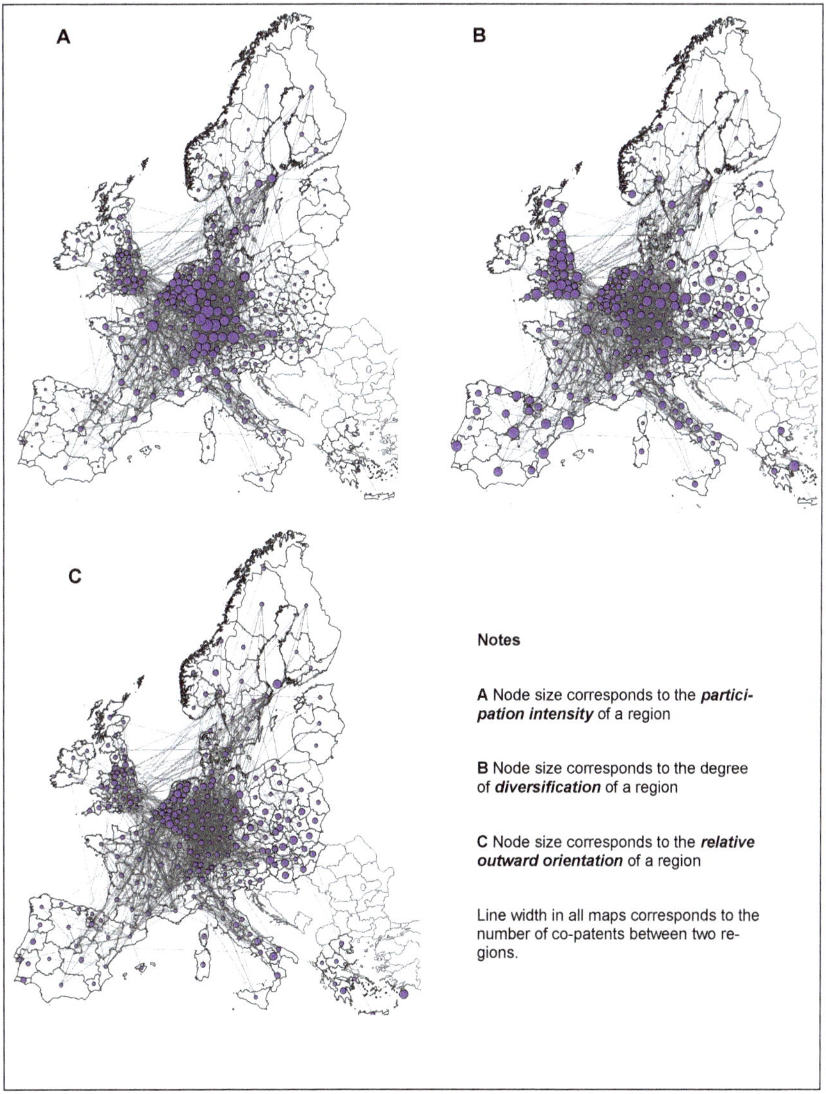

In addition, Figure 2 provides an illustrative overview in the form of spatial network maps of the European co-patent network, decomposed by the three components. The results are highly interesting, both in terms of illustrating the functioning of BC, as well as in terms of providing insights into the spatial dynamics of European co-patenting. It demonstrates why some regions, such as Ile de France, do not appear on top in terms of BC due to their lower relative outward orientation. Further, commenting on the overall picture, it can be seen that, we observe the classical regions of Ile de France as well as regions in Western Germany, the Netherlands and UK to come up with highest participation intensity, while Eastern and Southern European regions show a higher diversification of their links, i.e. actors in those regions seem to be not that selective in the choice of their partners as actors with a high reputation in European core regions. With respect to outward orientation, we also observe high values of some Southern and Eastern European regions. In terms of established co-patent links, they seem to be highly open, which could be explained by their reliance on external collaborations and knowledge sources, as well as the lack of internal collaboration structures.

4 Summary and conclusions

In the STI community, we can observe a lively debate on the notion of positioning indicators, shifting emphasis to actors in the innovation process and their R&D interactions with other actors. However, up to now only few indicators exist which are able to provide a rather comprehensive assessment of an actor's positioning in the innovation system, reflecting structural characteristics of its internal and external R&D linkages. In this study, we suggest a new approach for assessing the positioning of actors in innovation systems relying on the notion of bridging centrality (BC), inspired from the literature on complex networks. Based on the concept of bridging paths, i.e. a set of two links connecting three actors across three different aggregate nodes (e.g. organisations, regions or countries), we argue that triangulation in networks is a key issue for knowledge recombination and the extension of an actor's knowledge base.

As bridges are most often not empirically observable at the individual level of research teams, we propose an approximated BC measure that provides a flexible framework for dealing with the aggregation problem in positioning actors. Hereby, BC is viewed as a function of a node's (i) participation intensity in the network, (ii) openness to other nodes (i.e. the relative outward orientation of network links), and (iii) diversification of links to other nodes. With these three components – which are both intuitive and computationally simple – we provide an integrative perspective that enables us to achieve a better understanding of the role of certain actors in R&D networks.

An illustrative example on the co-patent network for European regions demonstrates the performance and usefulness of our BC measure for networks constructed at the aggregated level. Despite observing similar patterns in basic statistics like correlations of the centralities, we were able to show striking and interesting differences in the structure of the inter-regional co-patent linkages across regions. A region's outward orientation and the diversification of its network links moderate the influence of regional scale on network centrality. This is a major strength of the measure proposed in this study, and it paves the way for future studies to examine the role of certain aggregate nodes in R&D networks, not only among regions, but also among organisations, and, by this, contributes to the debate on positioning indicators in the STI context.

Of course, there is room for further improvements of the approach. Indeed, a crucial point for future research is to devote higher emphasis to the specific characteristics of R&D network links and our concept could be used to integrate these aspects. For example, extensions of the bridging centrality could include a focus on the fact that some bridging agents indirectly connect national actors with international ones. By focusing on technology related issues, one could consider bridging agents who indirectly connect actors from one specific technology with others from another technology. Moreover, the measure of bridging centrality is not limited to the context of R&D but may prove to be useful also for the application in other types of network structures, such as trade flows or economic value chains.

References

Bergé, L. (2016). Network proximity in the geography of research collaboration. *Papers in Regional Science* [forthcoming].

Berge, L., Wanzenböck, I. and Scherngell, T. (2015). Centrality of regions in R&D networks: Conceptual clarifications and a new measure, working paper Groupe de Recherche en Economie Théorique et Appliquée (Gretha), No. 2015-31.

Borgatti, S.P. (2005). Centrality and network flow. *Social Networks*, 27(1): 55–71.

Breschi, S. and Lenzi, C (2015). The Role of External Linkages and Gatekeepers for the Renewal and Expansion of US Cities Knowledge Base, 1990–2004. *Regional Studies*, 49(5): 782–797.

Burt, R.S. (1992). *Structural holes: The social structure of competition*. Harvard University Press, Cambridge, MA.

Edler, J., Fier, H. and Grimpe, C. (2011). International scientist mobility and the locus of knowledge and technology transfer. *Research Policy*, 40(6): 791–805.

Fischer, M.M., Scherngell, T. and Jansenberger, E.M. (2006). The geography of knowledge spillovers between high-technology firms in Europe. Evidence from a spatial interaction modelling perspective, *Geographical Analysis* 38(3), 288–309.

Glänzel, W. and Schubert, A. (2004). Analysing scientific networks through co-authorship. In *Handbook of quantitative science and technology research*, pages 257–276. Springer.

Heller-Schuh, B., Barber M.J., Henriques, L., Paier, M., Pontikakis, D., Scherngell, T., Veltri, G.A. and Weber, M. (2011). Analysis of Networks in European Framework Programmes (1984–2006). Sevilla, European Commission [ISBN: 9789279197147].

Lazega, E. and Snijders, T. (ed.) (2015). Multilevel network analysis for the social sciences. Springer.

Lepori, B., Barré, R. and Filliatreau, G., (2008). New perspectives and challenges for the design and production of s&t indicators. *Research Evaluation*, 17(1): 33–44.

Maraut, S., Dernis, H., Webb, C., Spiezia, V. and Guellec. D. (2008). The OECD REGPAT Database. *OECD Science, Technology and Industry Working Papers.*

OECD (2005). Oslo manual: Guidelines for collecting and interpreting innovation data. *Organisation for Economic Cooperation and Development (OECD), Paris.*

Robinson, W.S. (1950). Ecological correlations and the behavior of individuals. *American Sociological Review*, 15(3): 351–357.

Scherngell, T. (ed.) (2013). *The Geography of Networks and R&D Collaborations.* Springer- Physica Verlag, Berlin-Heidelberg-New York.

Scherngell, T. and Lata, R. (2013). Towards an integrated European Research Area? Findings from Eigenvector spatially filtered spatial interaction models using European Framework Programme data, *Papers in Regional Science* 92, 555–577.

Scherngell, T. and Barber, M. (2011). Distinct spatial characteristics of industrial and public research collaborations: Evidence from the 5th EU Framework Programme, *The Annals of Regional Science* 46(2), 247–266.

Scherngell, T. and Barber, M. (2009). Spatial interaction modelling of cross-region R&D collaborations. Empirical evidence from the 5th EU Framework Programme, *Papers in Regional Science* 88(3), 531–546.

Scherngell, T., Wanzenböck, I. and Berge, L. (2016). Bridging centrality: A new indicator to measure the positioning of actors in R&D networks. In: STI conference 2016 Valencia, 1106–1116

Snijders, T. (2016). The multiple flavours of multilevel issues for networks. In: *Multilevel Network Analysis for the Social Sciences*, pages 15–46. Springer.

Wanzenböck, I., Scherngell, T. and Brenner, T. (2014). Embeddedness of regions in European knowledge networks. A comparative analysis of inter-regional R&D collaborations, co-patents and co- publications, *The Annals of Regional Science* 53, 337–368.

Wanzenböck, I., Scherngell, T. and Lata, R. (2015). Embeddedness of European regions in EU funded R&D networks: A spatial econometric perspective, *Regional Studies* 49(10), 1685–1705.

Wasserman, S. and Faust, K. (1994). *Social network analysis: Methods and applications*. (Vol. 8). Cambridge university press.

Manfred Paier, Martina Dünser, Thomas Scherngell,
Simon Martin

Agent-based Modelling in RTI Policy: Application to the Austrian biotechnology industry

Abstract: *This chapter presents an empirical agent-based model of knowledge creation in a system of researching firms. The model focuses on government policies fostering inter-organisational collaboration, risk-taking propensity and competitive advantage. Simulations refer to the impact of such interventions on overall research output and technological specialization of the Austrian biotech industry.*

1 Introduction

Agent-based modelling (ABM) is a method for analysing complex systems with basically two features: first, the system is composed of interacting agents and second, the system exhibits emergent properties. ABM has been applied to fields as diverse as financial markets, human cooperation and scientific knowledge production (see, for instance, Edmonds et al. 2008). In economic applications, ABM provides a framework to simulate the behaviour and interactions of heterogeneous agents within and with a given environment, with the objective to represent the complexity of real-world systems (Hamill and Gilbert 2015).

This chapter presents an empirically based ABM in the field of Research, Technology and Innovation (RTI) policy, and provides scenario analyses of the impact of government programs in a system of innovating firms. The topic is highly relevant since governments are increasingly held accountable for their interventions, and they have to assess potential impacts before a program can be implemented. In 2002, the European Union officially introduced Ex-ante Impact Assessment (IA) as a compulsory procedure to analyse potential effects of new programs (Delanghe and Muldur 2007). While IA in general comprises a broad range of dimensions, our main interest is in the impact of policy intervention on knowledge creation in the innovation system.

An important debate in RTI policy is on technological specialization versus diversification and their effects on the innovative performance of an industry. It attained renascent importance through the policy concept of smart specialization (European Commission 2013). Beyond the discussion on whether actually specialization within an industry or rather diversity is more likely to promote innovative activities, we highlight the steering role of policy makers being in the position to actively change the technological profile of a region by e.g. fostering specialization through funding certain technologies more than others.

In our illustrative example, we focus on knowledge production in the Austrian biotechnology-related industry. Biotechnology is a knowledge intensive sector, characterised by high intensity of both research and knowledge exchange among economic actors (Owen-Smith and Powell 2004, Tödtling and Trippl 2007). In this case, economic performance of the industry is to a substantial degree driven by the generation of scientific knowledge and its direct application.

2 Methodology and model description

To master the extensive challenge of ex-ante evaluating the effects of policy interventions on knowledge creation in a system of interlinked firms, we apply an agent-based modelling approach. According to Macal and North (2010, p. 152), three key characteristics are inherent to ABMs:

- a set of agents, their attributes and behaviours,
- a set of relationships and methods of interaction,
- the agents' environment.

This approach facilitates a systemic, micro-based view of innovation and enables a tailored adjustment to specific policy interventions in different industries and time horizons. The behaviour of each single agent contributes to an aggregate system performance, which in turn influences the agents on an individual level. Regarding the ex-ante evaluation of policy instruments, this is an essential advantage, since such interdependencies are mostly difficult to assess beforehand and limited to questions linked to specific instruments.

The conceptual model of knowledge creation in biotechnology firms is derived from theoretical considerations and empirical observations in innovation economics. Core ideas are based upon the SKIN model (Simulating Knowledge dynamics in Innovation Networks), a multi-agent model containing heterogeneous agents which act in a complex and changing environment (Gilbert et al. 2001, Korber and Paier 2014, Triulzi et al. 2014).

The basic structure of our model can be subdivided into (i) an input side, with the agent's knowledge endowment, strategies and organisational figures, (ii) an interaction part with research processes in order to generate knowledge and (iii) an output side, where the knowledge output is realized in terms of knowledge gains and patents (see Figure 1).

Figure 2: Agents' processes (overview)

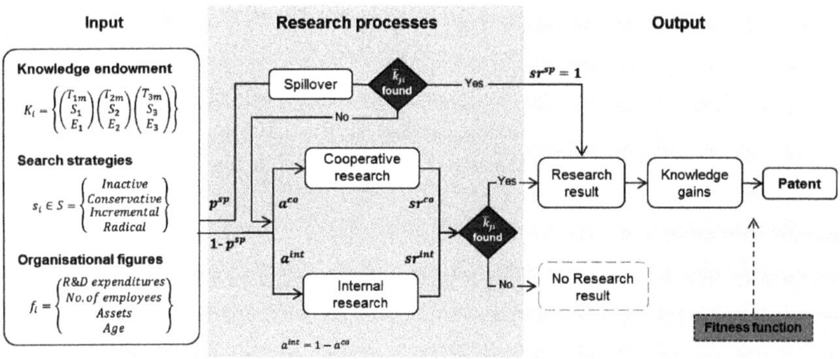

The relationship between these three parts is not linear, but is characterized by feedback-loops and interdependencies. The model of knowledge creation as a whole is embedded in a research system with industry-specific institutional characteristics, which is by itself affected by research policy interventions at various stages in the model (for details, see Paier et al. 2017).

3 Agents, attributes and strategies

We consider the agents in the model to be industrial firms that perform research. In order to do so, each agent is provided with a set of variables, i.e. agent-specific knowledge endowment, strategies and organisational figures. Thus, the agent population is heterogeneous and dynamic with respect to these attributes.

3.1 Knowledge endowment

The knowledge endowment of the agent represents the agent's dynamic knowledge base during the simulation. It depicts the technological field the agent is currently active in and upon which it may perform research. A subfield category is included, taking into account the fact that the agents may differ with respect to their core competencies, activities and operations. Moreover, an expertise level indicates how frequently and how long the agent has successfully conducted research in the research field and subfield. Formally, the knowledge endowment $K_i = \{k_{1i}, k_{2i}, k_{3i}\}$ of agent a_i $(i = 1, 2, ..., I)$ is defined as a set of three so-called kenes k_{ji}, similar to the concept of Gilbert (1997). The kene k_{ji} consists of a technology class T_{jm}, a subfield S_j and an expertise level E_j with $j = 1, 2, 3$ and $m = 1, 2, ..., M$. Hence, the knowledge endowment K_i of agent a_i is given by

$$K_i = \left\{ \begin{pmatrix} T_{1m} \\ S_1 \\ E_1 \end{pmatrix}, \begin{pmatrix} T_{2m} \\ S_2 \\ E_2 \end{pmatrix}, \begin{pmatrix} T_{3m} \\ S_3 \\ E_3 \end{pmatrix} \right\} \qquad (1)$$

where I and M denote the empirically determined numbers of agents and technology classes. The domains of the subfield $S_j \in \{1, 2, ..., 10\}$ and the expertise level $E_j \in \{1, 2, ..., 10\}$ are given by definition (i.e. by model design).

3.2 Strategies

The innovation system comprises a wide range of players with different strategies and behaviours. For instance, some firms are pure manufacturing facilities and hence do not engage in any research activities. On the contrary, some firms perform research intensively and play the role of highly specialized technology leaders in their traditional field. Other firms aim at diversifying into a broad range of research fields, either through incremental changes or, more riskily, through shifting to entirely different technology fields. This may enable firms to change to less competitive areas of research or spread their risk to reduce their vulnerability regarding external shocks.

To account for this heterogeneity, the knowledge creation model foresees two groups of agent strategies: The first group refers to choosing a research target, whereby an agent uses one out of four available *search strategies*. The second group, the agent's *research strategies*, relates to the possible ways how the chosen research target is striven for. Hence, each time step (i.e. a quarter of a year), an agent a_i engages in research activities by defining a research target $\bar{k}_{ji}$. For that purpose, the agent randomly chooses one of the three existing kenes k_{ji} and modifies it according to its given search strategy:

- ***Gridlock:*** With this strategy the agent does not perform any research and hence, no new research target is formed; the 'new' research target equals the old target:

$$\bar{k}_{ji} = k_{ji}$$

- ***Conservative*:** If the agent follows a conservative search strategy, it aims at increasing its expertise level E_j in a certain research field and subfield. Thus, its new research target is set to the following:

$$\bar{k}_{ji} = \begin{Bmatrix} \bar{T}_{jm} = T_{jm} \\ \bar{S}_j = S_j \\ \bar{E}_j = E_j + 1 \end{Bmatrix}$$

While the technology class and the subfield remain the same, the expertise level is increased by one. If the expertise level has reached its maximum level $E_j = 10$, the agent remains at this level of expertise.

- ***Incremental***: An agent with an incremental search strategy tries to modify its research orientation represented by a change of its subfield value S_j, while staying in the same technology class. Since a new area of research has opened up for the agent, the corresponding expertise level is set to a beginner's level $E_j = 1$:

$$\bar{k}_{ji} = \begin{Bmatrix} \bar{T}_{jm} = T_{jm} \\ \bar{S}_j = (S_j + 1)\ mod\ 10 \\ \bar{E}_j = 1 \end{Bmatrix}$$

  Every time the incremental strategy is applied, the subfield value is increased by one. If an agent with the maximum value $S_j = 10$ pursues the incremental strategy, its new subfield is set to $S_j = 1$.

- ***Radical:*** An agent with a radical search strategy goes for repositioning itself in a new technology class, associated with a diversification strategy. Thus, to define the new research target, a new technology class is chosen, whereby technological similarity is taken into account: Similar technology classes (to one of its current technology classes) are chosen more likely than distant ones[1]

$$\bar{k}_{ji} = \begin{Bmatrix} \bar{T}_{jm} \neq T_{jm} \\ \bar{S}_j = S_j \\ \bar{E}_j = 1 \end{Bmatrix}$$

The subfield of the old kene is inherited and hence remains unchanged. Again, due to the newly entered research field, the expertise level of the agent is set to one.

Once the research target is chosen, the research strategy defines the way how the agent tries to obtain it. The model provides three kinds of research strategies:

- ***Spillover***[2]: The agent receives the new kene if its research target $\bar{k}_{ji}$ is similar to another agent's kene k' in the population, i.e. if the technology class and subfield are identical and the expertise level fulfils the condition:

1 Note: $\bar{T}_{im}$ is chosen from the set of technology classes T based on an empirical similarity measure given by the Jaccard index of technology classes (see Subsection 4.1).

2 Strictly speaking, the notion of spillover is not a deliberate actor strategy since it describes a phenomenon that occurs unintentionally within a population. Nevertheless,

$$E(\bar{k}_{ji}) \leq E(k')$$

- ***Internal research:*** An agent tries to achieve its research target without depending on other agents.
- ***Cooperative research***: An agent looks for a partner with a cooperative strategy that holds an equal or similar kene k' as its research target $\bar{k}_{ji}$ set afore. The similarity of the desired kene depends on the agent's search strategy. Agents with a conservative search strategy only have a tolerance level δ^E regarding the expertise level,

  $$|E(\bar{k}_{ji}) - E(k')| \leq \delta^E$$

  whereas agents with an incremental or radical search strategy additionally have a tolerance level δ^S regarding the subfield:

  $$|S(\bar{k}_{ji}) - S(k')| \leq \delta^S$$

3.3 Organisational figures

In addition to the knowledge endowment and the strategies, each agent is individually characterised by organisational figures, namely research expenditures R_i, number of employees L_i, assets I_i and age A_i. These figures, representing the fitness of an agent, are essential for the evaluation of the agent's research success in the output part.

3.4 Research processes and system output

The process of knowledge creation on the agent level links the input and the output part of the model as illustrated in Figure 1. An agent starts with a certain probability p^{sp} that it may receive knowledge through spillover. This is based on the assumption that a certain percentage of knowledge is non-excludable and a non-rival public good (Fischer and Fröhlich 2001). If the agent finds an appropriate kene for matching its research target through spillover, the research result is taken for granted, which ends up with an achieved research result and thus completes the agent's process of knowledge creation. Alternatively, i.e. with the probability of $1 - p^{sp}$, the agent engages either in cooperative research (with probability a^{co}) or in internal research (with probability $a^{int} = 1 - a^{co}$). In case of a missing match from spillover, the agent engages in research according to these fractions as well. Now, the attainment of the research result depends on the success rates sr^{co} and

in the model, the process can be formally conceptualised in the same procedure as the two other research strategies.

sr^{int}. If the research result is actually achieved through these research processes, the kene $\bar{k}_{ji}$ replaces the old one k_{ji} in the knowledge endowment K_i of the agent.

In the output part of the model, successfully achieved research results are incorporated into the agent's knowledge endowment and as such represent knowledge gains as described above. Additionally, the agent undergoes a fitness test which determines the agent-specific patenting propensity using empirical evidence as described in more detail in Subsection 4.3. This fitness function represents a filter that determines whether the agent's knowledge gains classify for becoming a patent.

4 Introducing empirical data to the model

In order to adapt the model to the specific case of the Austrian biotechnology-related industry, we follow an empirical agent-based Modelling approach (Smajgl and Barreteau 2014). Apart from the use of expert knowledge during model conceptualization and scenario formation, this section describes how empirical data is introduced into the model to (i) initialize the agent population (Input), (ii) to calibrate the model (Research processes) and (iii) to determine the research output (Output).

4.1 Initialization of the agent population

The model is initialized at the agent level as well as at the system level. For initialization at the agent level, we include 61 private firms of the Austrian biotechnology-related industry. The sample is composed of firms with active patenting records in patent classes associated with biotechnology (OECD 2008) over the years 2000–2010, extracted from the PATSTAT database. The set of technology classes individually assigned to each agent corresponds to the patent classes in which they actually hold patents in the period of initialization (i.e. 2000–2010). The patent classes are used on a three-digit subclass level specified by the International Patent Classification (IPC). The number of technology classes in the model results from the most frequently occurring IPC classes on the patents of the firm sample (i.e. 45 technology classes). Additionally, 16 other related patent classes are considered[3] to provide the possibility for further knowledge creation.

For the initialization of an agent's knowledge endowment, first, for every kene k_{ji}, a technology class is randomly drawn from the set of references to IPC classes

3 Note that the identical number of total technology classes and agents (i.e. 61) is not intentional but stems from the empirical initialisation of the model.

(from the associated firm's individual patent stock). Second and third, the respective subfield and expertise level are chosen randomly according to the specifications stated in Eq. (1). Moreover, each agent a_i is also individually equipped with four empirically based organisational figures: (i) research expenditures R_i, (ii) number of employees L_i, (iii) assets I_i and (iv) age A_i, taken from a company database and a recent industry study (Schibany et al. 2010, Orbis 2014).

At the system-level, an important aspect of model initialization is the notion of an empirically based "knowledge space" in which the agents intentionally move around if they achieve new knowledge. To this end, a relational concept is needed that determines the technological distance between the technology classes. This can be obtained using a similarity measure on the set T of technology classes derived from the (empirical) patent stock of the firm sample. Hereby, we define the similarity of two technology classes T_l and T_m (with $l, m = 1, 2, ..., M$) as the Jaccard index J_{lm} (Rip and Courtial 1984) given by

$$J_{lm} = \frac{c_{lm}}{c_l + c_m - c_{lm}}$$

where c_{lm} denotes the number of co-references to technology classes T_l and T_m, while c_l and c_m denote the numbers of references to technology classes T_l and T_m in the given set of patents, so that $0 \leq J_{lm} \leq 1$.[4] By means of this definition, two technology classes are considered more similar the more often they are both mentioned in a patent. In the model, we assume that a new technology class is more easily accessible for an agent if it is more similar to its existing knowledge endowment.

4.2 Calibration of the research processes

Unlike the agent characteristics, which are initialized with empirical firm data, a set of free system parameters governs the model processes. To calibrate the model, these parameters have to be adjusted in such a way that the simulation output renders the empirical data of the real-world reference system (see Table 1).

The particular empirical focus of our modelling approach imposes empirically justified restrictions on the calibration process. To this end, a two-step calibration strategy is followed, called *fractional factorial design* (Thiele et al. 2014). In the first calibration step, representing the empirical restriction, a range of possible values for each parameter is defined based on expert knowledge. In the second

4 The empirical values of J_{lm} used in the current application (61-by-61 matrix) can be obtained from the authors.

calibration step, the search for the best fit between simulated and observed output data is performed by systematic parameter sweeping in the preselected part of the parameter space. In our case, two patent-related quantities are chosen as matching criteria between empirical and simulated data: (i) the total number of patents in the firm population and (ii) the patenting profile of the population, i.e. the distribution of these patents over the patent classes.

The empirical reference dataset is the patent performance of Austrian biotechnology firms in the five-year period of 2008–2012, and the calibration is performed with the simulated patent output after 20 time-steps (four time-steps representing one year). The best fitting parameter values are chosen in such a way that the empirical and simulated patents exactly match in terms of their total number, and reach the highest possible similarity with respect to their profiles, whereby the similarity of the profiles is measured as the Pearson correlation coefficient of the corresponding vectors in the knowledge space.

Table 1: Calibrated system parameters

Parameter	**Description**	**Calibrated value**
Gridlock (a^{gri})	Agent share with no research	0.5
Conservative (a^{con})	Agent share with conservative search strategy	0
Incremental (a^{inc})	Agent share with incremental search strategy	0
Radical (a^{rad})	Agent share with radical search strategy	0.5
Search dispersion (r^{sd})	Search radius for technology classes	0.9
Spillover (p^{sp})	Probability of local knowledge spillover	0.5
Coop (a^{co})	Share of agents conducting cooperative research	0.6
Internal (a^{int})	Share of agents conducting internal research	0.4
Success rate coop (sr^{co})	Probability of successful cooperative research	0.6
Success rate internal (sr^{int})	Probability of successful internal research	0.4
Patenting rate (p_0)	Base patent probability	0.4

Note: By definition, the parameters referring to population shares have to fulfil the constraints $a^{gri} + a^{con} + a^{inc} + a^{rad} = 1$ and $a^{co} + a^{int} = 1$.

Note that by choosing this empirical calibration strategy, one implicitly refers to the actual institutional framework conditions the Austrian biotechnology industry was embedded in, including the latent R&D policy during the calibration period. Thus, this parameter set defines the baseline scenario, as a reference for the simulations presented in Section 3.

4.3 Evaluation of the agents' research results

As stated in Subsection 3.4, the model comprises an empirical output filter determining if an agent's achieved knowledge gain gives rise to a patent (also see Figure 1). The probability that the knowledge gain of an agent becomes a patent depends on a system parameter p_0 (the base patenting probability) and an agent-specific patenting propensity, an empirically determined fitness function f_i depending on its organisational figures:

$$p_{pat_i} = p_0 f_i\,(R_i, L_i, I_i, A_i) \tag{2}$$

with $$f_i \sim \exp(\beta_1 R_i + \beta_2 L_i + \beta_3 I_i + \beta_4 A_i)$$

where R_i denotes the research expenditures, L_i the number of employees, I_i the assets, and A_i indicates the age of agent a_i. The coefficients β_1, β_2, β_3 and β_4 are estimated by means of a zero-inflated negative binomial model for a sample of 155 patenting and non-patenting Austrian biotechnology firms. In this estimation procedure, the firms' organisational figures are set in relation to the observed number of patents:

$$Y_i = f(X_i) \tag{3}$$

where $X_i = (R_i, L_i, I_i, A_i)$. Since the dependent variable, namely the number of patents, is count data, the Poisson regression framework is an appropriate approach (Cameron and Trivedi 2012). The employed Poisson regression model is specified as follows:

$$\mathrm{Prob}(Y = y_i) = \frac{e^{-\lambda_i}\lambda_i{}^{y_i}}{y_i!}, \tag{4}$$

where $\ln\lambda_i = \beta' X_i$ where λ_i is the conditional mean, and β is the k-by-1 vector of parameter estimates with $y_i = 0, 1, 2, \ldots$ as well as $i = 0, 1, 2, \ldots, I$.

Note that the Poisson model from Eq. (4) is based on the assumption of equidispersion, i.e. equality of conditional mean and variance. This assumption is no longer valid if there is strong heterogeneity among the observations (i.e. overdispersion). To cope with this circumstance, we apply an extension of the Poisson regression model, the negative binomial regression model. The variance is now assumed to be a quadratic function of the mean: $\lambda_i + \alpha\lambda_i^2$ (Cameron and Trivedi 2012). This yields $\ln\lambda_i = \beta' X_i + \varepsilon$ with $\exp(\varepsilon)$ following a gamma distribution. Hence, the probability distribution of (4) becomes

$$\text{Prob}(Y = y_i|\varepsilon) = \frac{e^{-\lambda_i \exp(\varepsilon)} {\lambda_i}^{y_i}}{y_i!} \tag{5}$$

Transforming (5) to an unconditional distribution of y_i by integrating ε out of the expression yields:

$$\text{Prob}(Y = y_i) = \frac{\Gamma(\theta + y_i)}{\Gamma(\theta) y_i!\, u_i\, (1-u_i)^{y_i}}, \tag{6}$$

where $u_i = \frac{\theta}{(\theta+\lambda_i)}$ with $\theta = \frac{1}{\alpha}$. The additional parameter α is referred to as dispersion parameter. Since many firms in the sample had no patenting activities, the data contains a relatively high number of zeroes. Hence, a zero-inflated model is applied, where the actual model is preceded with a binary logit model.

Two estimates of the regression model show significant effects, namely the number of employees and the R&D expenditures. In our empirical context, the number of employees has a negative effect with a coefficient of -0.020, which suggests that smaller biotechnology firms have a higher patenting propensity. This finding can be explained by the structure of the Austrian biotechnology-related industry, where highly specialized and successful small and medium sized firms are active in specific niches. Not surprisingly, R&D expenditures have a positive effect on the knowledge output in the form of patents. The two remaining regressors show neither critical values nor are they significant. The Vuong test to decide between the zero-inflated and the standard negative binomial model is statistically significant, suggesting the use of a zero-inflated model (Long and Freese 2001). The resulting coefficients of the regression are used as fixed parameter values in the model and hence, determine the patenting probability of the agents through the fitness function stated in (2).

5 Policy scenarios and simulation results

5.1 Policy scenarios

By varying system parameters, policies that deviate from the business-as-usual policy (i.e. the baseline scenario) can be modelled. It is important to note that the direct effect of the policy intervention on the individual agent is assumed to be exogenous, and the endogenous changes of the agents' knowledge endowment stem from the structure of the technology space and the agents' interactions. The focus of the model is on the long-term evolution of the knowledge base of the firm population under different policy interventions. We refer to the baseline scenario defined in the previous subsection and interpret it as the result of an empirically given aggregate policy without specification of the programs that have been in place during the reference period. Thus, the baseline scenario reflects a business-as-usual policy. In contrast, the following policy scenarios are conceived as alternative government interventions.

5.2 Policies focusing on inter-organisational collaboration

Funding cooperative research is a common policy intervention to create positive external effects, for instance knowledge spillover or dynamic effects of knowledge creation. At the system level, advantages arise from synergies, the transfer of knowledge and the resulting knowledge gains of all cooperation partners. However, collaboration is not a goal per se, and an ongoing policy discussion is driven by the issue of complementarity between individual and inter-organisational research (e.g., Cunningham and Gök 2015). With our model we compare the long-term effect of two policies that effectuate different degrees of research collaboration.

Policy focus on cooperative research

The creation of new knowledge and its diffusion through inter-firm research cooperation is the main argument for this type of policy intervention (Hoekman et al. 2009, Reinold et al. 2013) to represent a policy focus that supports inter-organisational research collaboration. In terms of system parameter values, this scenario is characterized by an increased share of cooperative agents a^{co}, a higher probability of spillover p^{sp}, as well as a higher success rate sr^{co} for cooperative research (see Table 1 for parameters).

Policy focus on individual research

The reasoning of funding individual research is to avoid potential free riding and shirking problems that may occur in collaborative research. Besides, considerable additional organisational work and expenses may be avoided, accelerating individual research (Klette et al. 2000). Furthermore, the division of individual property rights is not an issue. In order to represent a higher level of individual research, two parameter values are changed in comparison to the baseline scenario: the share of individually researching agents a^{int} is increased (associated with a decreased share of cooperatively researching agents a^{co}) and a reduced probability of spillover p^{sp}.

5.3 Policies focusing on risk-taking propensity

Due to the imperfect appropriability of new knowledge, private returns to R&D are assumed to be lower than its social returns. That is why governments commit funds for stimulating business research and try to reduce the risk of private R&D (associated with both the cost and uncertainty of research). However, the effectiveness of government subsidies can be challenged on three main grounds: crowding out, substitution effects and allocative distortions (Guellec and van Pottelsberghe 2003). An important policy debate is therefore on the unintended effects of direct financial support (government grants) or indirect funding (e.g. tax reductions) for private firms. In our model, risk-taking is related to technological uncertainty firms are facing when they try to enter or master a technological area that is new to the firm.

Policy focus on direct funding

Direct funding subsumes the direct transfer of public financial resources to finance research projects (government grants). Due to the asserted substantial takeover of risk by the government in this case, the agents are more inclined to explore new and also more "distant" technological areas. Hereby, direct funding is assumed to be "bottom-up" funding, i.e. the direction of the research is defined by the firm strategy alone and there is no technological restriction as in "top-down" funding programs (Astor et al. 2009). In the model, the direct funding policy is represented by increasing the share of agents with radical search strategy (increased system parameter a^{rad}), and an expanded search radius in technology space (increased system parameter r^{sd}) as compared with the baseline scenario.

Policy focus on indirect funding

In contrast to direct monetary grants, indirect funding takes the form of easing the tax burden by refunding a certain amount of tax payment if a research is conducted, independent of the research area or research success. In this case, the risk is also lowered to some degree, but the assumption is that risk reduction is not as pervasive as in the case of receiving subsidies via direct funding (Mohnen and Lokshin 2009). In the model, this policy is defined by reducing the share of radical agents a^{rad}, whereas the shares of conservative a^{con} and incremental research agents

a^{inc} are increased. It follows that relatively more agents either increase only their expertise level or move to a new subfield, but do not change the technology class.

5.4 Policies focusing on competitive advantage

Policy interventions targeting technological competitiveness either try to balance out technological weaknesses or support existing strengths. Whereas the first is intended for building up new or strengthening less advanced technologies (Park and Lee 2006), the latter aims at expanding the technology lead in specialized research fields, establishing a comparative advantage (Cantwell and Iammarino 2001), a strategy that is also reflected in the concept of smart specialization, prioritizing research fields where already a cluster of activities is present and well established (Foray et al. 2011).

Policy focus on overcoming weaknesses

In the simulation model, the focus on overcoming weaknesses regarding the innovative performance in certain research fields is interpreted as increased funding of small research fields, meaning those where not many firms showed patenting activity. This is done by assigning these technology classes higher probabilities to be chosen as research targets by the agents. The system parameters in Table 1 remain unchanged.

Policy focus on advancing the strengths

Contrary to the funding of small research fields, the idea of constructing regional advantage by means of increased funding for already dominating research fields (in terms of patenting activity) is implemented by increased probabilities for the six largest technology classes to be chosen by the agents to pursue their research. Again, the model parameters are kept the same as in the baseline scenario (according to the parameter values resulting from the calibration).

Policy focus on thematic clusters

In order to simulate a funding focus on clusters with thematically related research fields, exemplary, a group of technology fields in the areas of biochemistry, microbiology and enzymology are selected. These fields correspond to sub-classes of the patent class C12 and can hence be seen as thematically close. Again, the funding focus is translated so that agents – while defining their new research target – have higher probabilities to choose the technology fields within this cluster.

6 Simulation results

The scenarios are compared and interpreted with regards to their number of patents (cumulated over 30 years) and their degree of specialization, measured in terms of the Krugman specialization index.[5] In this way different effects can be observed and analysed:

1. Effect of certain policy interventions on
   a. the research output
   b. the degree of specialization

2. Correlation between research output and degree of specialization

Figure 2 presents an overview of the simulation results, by comparing the policies with respect to output quantity and specialization[6]. The number of patents is positively correlated with the degree of specialization. However, the policy scenario of focusing on direct funding stands out in this respect. It shows by far the highest number of patents but shows a fairly strong diversification among the technology classes, since direct funding in the model seems to especially promote the "smaller" research fields. This results from the increased number of radical agents and their expanded search horizons while choosing new technology classes as their research targets. Relatively high degrees of specialization as well as high numbers of patents are observed with the policy scenarios of indirect funding, focusing on cooperative research and increased funding of dominating research fields.

Focus on indirect funding seems to only reach higher numbers of patents in a few large technology classes. This concentration on "larger" classes is due to the reduced share of agents with radical search strategy. From an innovation policy perspective, this is by all means plausible, since for example, the indirect funding scenario may represent tax reduction and exhibits a characteristic phenomenon – the free-rider-effect. In this case all agents are favoured, no matter if research would have been conducted anyway. This may indicate a reduced effectiveness of this funding type, since no significant changes in behaviour can be observed: Higher numbers in patenting only occur in already predominant technology classes and there is little diversification to new technology classes.

5 The Krugman specialization index (KI) is defined as: $KI = \sum_{m=1}^{M} \left| b_m - \bar{b}_m \right|$, with $b_m = e_m / \sum_{m=1}^{M} e_m$ where e_m denotes the number of patents per research field.

6 A more detailed account of the individual scenario results, together with a specification of the scenario-specific exogenous parameters, is given in Paier et al. 2017.

A policy focus on cooperative research results in a comparably higher patenting activity due to a relatively higher probability of knowledge spillover and a higher success rate of cooperative research. Counterintuitively, the increased number of cooperatively researching agents dampens the knowledge growth in the first place. This is caused by the necessity of finding a suitable cooperation partner in order to conduct research; however, this partner is not always available in the population. The high degree of specialization within this scenario is the self-enhancing effect of agents increasingly looking for similar partners to conduct cooperative research. The more agents are active in a particular technology class, the higher is the probability to find suitable partners. From a policy perspective, these results do not come unexpectedly: if a policy succeeds in increasing knowledge exchange through cooperative research, spillover and mutual learning should have concentration effects in technology fields where there is already a high level of activity.

Increased funding of dominating research fields leads to a strong specialization of the technological profile. It promotes the overall research output; however, patenting activities in the smaller, non-targeted technology classes is almost non-existent. From a policy perspective, this is a striking result: with a policy focus on specialization, like it is the case with the Smart Specialization and Comparative Advantage concepts, smaller research fields, such as niches or possible future technologies would be given no room to evolve.

A special case of focusing on dominating technology classes is the funding of prosperous thematic clusters. The additional thematic relatedness leads to self-enhancing effects among the related classes, leading to an overall increased patenting activity compared to the baseline scenario. However, positive effects on the remaining classes fall short.

The opposite of focusing on dominating research fields, is the policy scenario of funding predominantly smaller technology fields with so far only little patenting activity. This scenario shows the smallest number of cumulated patents as well as the smallest degree of specialization. Still, due to the diversifying character, this policy intervention may be to some extent attractive for policy makers, since it spreads the risk onto different, not necessarily thematically similar, research fields.

Promoting individual, in-house research also results in a relatively low degree of specialization and only a low number of overall patents. Here, small technology classes benefit, resulting in patents being more diversified among the technology classes. This follows from supporting independently of their technology field and suitable collaboration partners.

Figure 3: Summary of simulation results

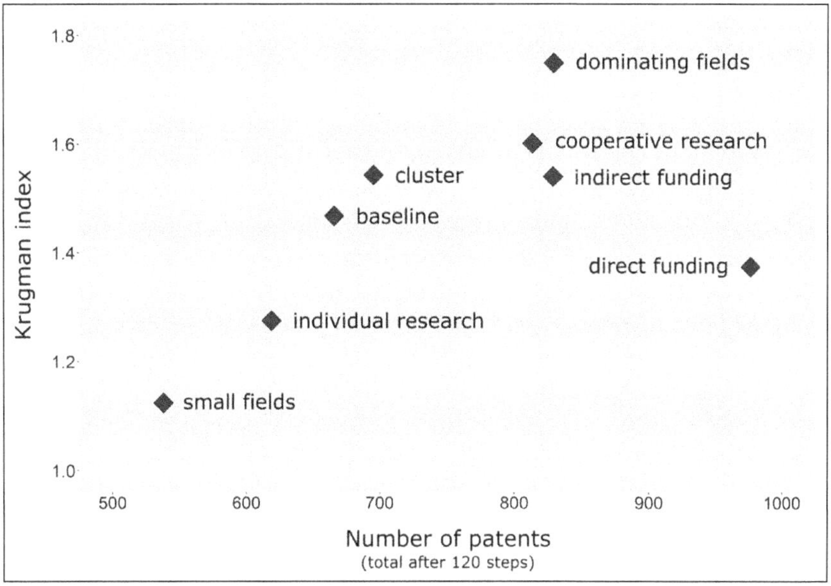

A comparison of different policy interventions, points out the steering role of policy makers with respect to specialization of regional technological profiles and the research output. Often, benefits due to specialization within a particular industry and due to regional concentration are seen as main drivers for innovative activities and the growth of industries (Marshall-Arrow-Romer specialization externalities). However, the importance of the diversity of industries and complementarity of knowledge of the agents for innovation and growth is not neglectable (Jacobs' diversification externalities). On the one hand, technological specialization may reveal comparative advantages and hence, create competitiveness in selected research areas. On the other hand, at a certain level, specialization may hinder knowledge generation and therefore innovativeness, due to lock-in effects (Feldman and Audretsch 1999). In this respect, fostering diversity is one way to reduce the risk of shocks cause by disruptive changes in technologies.

By inducing different policy interventions, it lies upon the political decision maker to balance between the degree of specialization or diversification of the technological portfolio and the amount of research output.

7 Summary and concluding remarks

In this chapter, we present an empirical agent-based model of knowledge creation in the Austrian biotechnology-related industry and simulate effects of different modes of public research funding on knowledge-related system output. Agent-based modelling is a methodological approach particularly suitable for describing and analysing complex systems consisting of interdependent actors, each of them acting on an endowed rule set. With our model we simulate the long-term evolution of the knowledge base of a given heterogeneous firm population given their research strategies and the structure of the relevant technological field. The relevance of this approach is in advancing the method as well as in contributing to the application of simulation in the context of policy analysis. The simulations refer to government policies that differ from the empirically calibrated baseline scenario in different aspects:

First, we compare two policy alternatives with a stronger focus on inter-organisational collaboration, fostering more collaboration or more individual research activities, respectively. The most important findings in this respect are that, more collaboration increases total patent output and concentrates the patent profile so that already strong technology classes become even more dominant. On the other hand, strengthening individual research dampens total patenting output, but gives rise to a broader diversification in the patent profile of the firm population.

Second, we simulate two alternative policies focusing on the risk-taking propensity by either reducing technological risk through direct research funding (government grants) or by reducing the cost of research while eventually discouraging risk-taking through indirect funding (tax reductions). The results of the simulation reveal that direct funding gives rise to a broadening of the patent profile, especially to the emergence of new, related technology clusters. A stronger focus on indirect funding, on the other hand, is able to increase the total number of patents, but conserves the existing technology profile. These results highlight the potential of applying the model within the context of ex-ante evaluation of policy measures.

Third, we analyse a triple of policy scenarios regarding the funding focus on weaknesses and strengths of the knowledge portfolio: focus on weaknesses (small technology classes), focus on strengths (dominating technology classes) and focus on prosperous thematic clusters (thematically related technology fields). The results show an increased overall patenting activity; however, this activity is strongly concentrated on the funded technology classes, whereas the small classes nearly diminished. Promoting the small classes on the other hand diversifies the technological profile, but shows only a low number of patents. Funding of thematic clusters clearly shows the effect of the relatedness of the research field, since they mutually influence each other in terms of their patenting activity, implicating an increase in specialization.

The presented results on the biotechnology industry underline the viability of the approach to simulate potential policy interventions on the knowledge base of a sectoral innovation system. At this stage of model development, simplicity and transparency have been the guidelines. Thus, the focus is on the complexities of knowledge creation only, while more market-related aspects are deliberately kept simple. Therefore, the interpretation of the current model is to be confined to the knowledge profile of the population, i.e. on knowledge creation at the system level. Future work will introduce additional complexity to the model with respect to exploitation of the knowledge and population dynamics, broadening the applicability to other industries. This will further increase the credibility of ABM for ex-ante impact assessment of RTI policy.

References

Cameron AC, Trivedi PK (2012). Regression analysis of count data. Cambridge University Press, Cambridge, UK.

Cantwell J, Iammarino S. (2001). EU regions and multinational corporations: Change, stability and strengthening of technological comparative advantages. Industrial and Corporate Change 10(4): 1007–1037.

Cunningham P, Gök A (2015). The Impact of Innovation Policy Schemes for Collaboration. In: Edler J, Cunningham P, Gök A, Shapira P (eds.) Handbook of Innovation Policy Impact. Edward Elgar Publishing Limited, London.

Delanghe H, Muldur U (2007). Ex-ante impact assessment of research programmes: the experience of the European Union's 7th Framework Programme. Science and Public Policy 34(3): 169–183.

Edmonds B, Hernández C., Troitzsch K. G. (eds.) (2008). Social Simulation: Technologies, Advances, and New Discoveries. Hershey, NY: Information Science Reference (IGI Global).

European Commission (2013). Smart specialization platform. URL:.http://s3platform.jrc.ec.europa.eu/home;jsessionid=CF5PSJRQnkR5G0K rb2jpvf L7bMcLWLv24pGsrVQFrF7yQHCNnRY9!948003041!1380553072323 (accessed 16 September 2013).

Feldman MP, Audretsch DB (1999). Innovation in cities: Science-based diversity, specialization and localized competition. European Economic Review 43(2): 409–429.

Fischer MM, Fröhlich J (2001). Knowledge, Complexity and Innovation Systems: Prologue. In: Fischer MM, Fröhlich J (eds.) Knowledge, Complexity and Innovation Systems. Springer, Berlin, Heidelberg, New York.

Gilbert N (1997). A Simulation of the Structure of Academic Science. Sociological Research Online 2(2):http://www.socresonline.org.uk/socresonline/2/2/3.html.

Gilbert N, Pyka A, Ahrweiler P (2001). Innovation Networks – A Simulation Approach. Journal of Artificial Societies and Social Simulation 4(3): 8.

Guellec D, van Pottelsberghe B (2003). The Impact of Public R&D Expenditure on Business R&D. Economics of Innovation and New Technology 12(3): 225–243.

Hamill L, Gilbert N (2015). Why agent-based modelling is useful for economists. Agent-Based Modelling in Economics: John Wiley & Sons Ltd, 1–10.

Hoekman J, Frenken K, van Oort F (2009). The geography of collaborative knowledge production in Europe. The Annals of Regional Science 43(3): 721–738(18).

Klette TJ, Møen J, Griliches Z (2000). Do subsidies to commercial R&D reduce market failures? Microeconometric evaluation studies. Research Policy 29(4–5): 471–495.

Korber M, Paier M (2014). Simulating the Effects of Public Funding on Research in Life Sciences: Direct Research Funds vs. Tax Incentives. In: Gilbert N, Ahrweiler P, Pyka A (eds.) Simulating the Knowledge Dynamics of Innovation Networks. Springer, Berlin Heidelberg New York.

Linton K, Stone P, Wise J (2008). Patenting Trends and Innovation in Industrial Biotechnology. In: *Staff Research Study 31*. U.S. International Trade Commission, Office of Industries, Washington, DC.

Long JS, Freese J (2001). Regression models for categorical dependent variables using STATA. Stata Corporation, College Station, TX.

Macal, C. M., & North, M. J. (2010). Tutorial on agent-based modelling and simulation. *Journal of simulation*, *4*(3): 151–162.

Mohnen P, Lokshin B (2009). What Does it Take for an R&D Tax Incentive Policy to Be Effective? http://d.repec.org/n?u=RePEc:cir:cirwor:2009s-11&r=ino, CIRANO Scientific Series, Montreal.

OECD (2009). The Bioeconomy to 2030: Designing a Policy Agenda. Main findings and policy conclusions. OECD International Futures Project, OECD, Paris.

OECD (2008). OECD Patent Databases: Identifying Technology Areas for Patents. OECD Publishing, Paris. http://www.oecd.org/sti/inno/40807441.pdf.

OECD (2015). Statistical Definition of Biotechnology. http://www.oecd.org/sti/biotech/statisticaldefinitionofbiotechnology.htm, last accessed 22 January 2016.

Orbis (2014). ORBIS company database. Bureau van Dijk.

Owen-Smith J, Powell WW (2004). Knowledge Networks as Channels and Conduits: The Effects of Spillovers in the Boston Biotechnology Community. Organisation Science 15(1): 5–21.

Paier M, Dünser M, Scherngell T, Martin S (2017). Knowledge Creation and Research Policy in Science-Based Industries: An Empirical Agent-Based Model. In Vermeulen B, Paier M (Eds) Innovation Networks for Regional Development. Concepts, Case studies, and Agent-Based Models: Springer International Publishing Switzerland, 153–183.

Park KH, Lee K (2006). Linking the technological regime to the technological catch-up: analyzing Korea and Taiwan using the US patent data. Industrial and Corporate Change 15(4): 715–753.

Reinold F, Paier M, Fischer MM (2013). Joint knowledge production in European R&D networks: Results from a discrete choice modelling perspective. In: Scherngell T (ed.) The geography of networks and R&D collaborations. Springer, Berlin Heidelberg New York.

Rip A, Courtial P (1984). Co-word maps of biotechnology: An example of cognitive scientometrics. Scientometrics 6: 381–400.

Schibany A, Berger M, Buchinger E, Dachs B, Dinges M, Ecker B, Falk M, Gassler H, Heller-Schuh B, Hofer R, Huber P, Janger J, Reinstaller A, Streicher G, Unterlass F (2010). Österreichischer Forschungs- und Technologiebericht 2010. In *Lagebericht gem. § 8 (1) FOG über die aus Bundesmitteln geförderte Forschung, Technologie und Innovation in Österreich*. BMWF, BMVIT, BMWFJ, Wien.

Smajgl A, Barreteau O (2014). Designing Empirical Agent-Based Models: An Issue of Matching Data, Technical Requirements and Stakeholders Expectations. In: Smajgl A, Barreteau O (Eds) Empirical Agent-Based Modelling – Challenges and Solutions. Volume 1: The Characterisation and Parameterisation of Empirical Agent-Based Models. Springer, New York Heidelberg Dordrecht London.

Thiele JC, Kurth W, Grimm V (2014). Facilitating Parameter Estimation and Sensitivity Analysis of Agent-Based Models: A Cookbook Using NetLogo and 'R'. Journal of Artificial Societies and Social Simulation 17(3): 11.

Tödtling F, Trippl M (2007). Knowledge Links in High-Technology Industries: Markets, Networks, or Milieu? The Case of the Vienna Biotechnology Cluster. International Journal of Entrepreneurship and Innovation Management 2/3/4/5: 345–365.

Trippl M, von Gabain J, Tödtling F (2006). Policy agents as catalysts of knowledge links in the biotechnology sector. SRE-Discussion 2006/01, Institute for Multi-level Governance and Development, Department of Socioeconomics, Vienna University of Economics and Business.

Triulzi G, Pyka A, Scholz R (2014). R&D and knowledge dynamics in university-industry relationships in biotech and pharmaceuticals: an agent-based model. International Journal of Biotechnology 6 13(1–3): 137–179.

Edgar Schiebel

Identifying research fronts: A science mapping exercise for security research

Abstract: *Bibliometric models of science dynamics cover several dimensions of a complex system: the history of the knowledge producing system, the current emergence of knowledge and interaction of actors in co-author networks. The AIT-Software BibTechMon is a tool to fulfil such tasks with a science mapping approach. We introduce the reduction in science maps coming from relevant publications and apply the idea in the case of security research.*

1 Introduction and methodology

In the last two decades, security issues in technology, economy, and society have become more and more important. The European Union is faced with new and changing opportunities and threats. In today's global society, risks and vulnerabilities are more diverse and less visible because political, social and technological developments have created a fluid security environment. The current global crisis regarding the refugee flows reflects several security challenges: politics, food security, etc. Already 10 years ago – without losing actuality – the European Commission (2007) reported to the European Parliament and the Council about the need to protect our critical infrastructure like roads, railways, bridges, information and communication infrastructures and power stations. In this chapter, current research fronts in security relevant issues are investigated with a focus on critical technologies and so-called soft issues like international security, food safety, information, communication technology and others.[1] Relevant research fronts are mapped with the help of a bibliographic coupling approach.

---

1 This work refers to work done in the project ETCETERA (2013), financed by the European Commission. The ETCETERA project was a contribution to efficient European security research planning. The project took up the twofold structure of topic "SEC-2010.7.0–3 Critical and emerging technologies for security" by dealing with the issues "Critical Technologies" and "Emerging Technologies" in two separate but interrelated research strands.

Bibliographic coupling and co-citation together with textual approaches are of growing interest to identify content-based subsets of publications in research fields. Price (1965), Kessler (1965), as well as Chen & Morris (2003) published pioneering work about the identification of research fronts by bibliographic coupling and knowledge bases by co-citation analysis. Boyack, K. and Klavans, R. (2010), Glänzel, W. (2012) contributed to a remarkable advance after this first pioneering work by examining the performance of different bibliometric approaches to delineate and identify issues in broader research fields. In recent years, traditional bibliometric concepts like co-citation analysis and bibliographic coupling have been enriched with text and semantic similarities, topic modelling (Yau, C., K., Porter, A., Newman, N., Suominen, A. 2014), as well as hybrid methods (Janssens, F., Glänzel, W., & Moor, B. 2008), using a combination of textual and bibliographic coupling. Additionally visualization tools for networks and the representation of agglomerations of similar documents in a two dimensional space (Van Eck, N.J. and Waltman, L. 2010; Chen, C. 2006, Kopcsa, A. and Schiebel, E. 1998), interactive visualizations of hierarchical cluster analysis and word clouds are helpful instruments that have been developed and are now broadly applied in science mapping exercises. Applied algorithms are cluster analysis, multidimensional scaling and spring models. Recently latent Dirichlet allocation for topic modelling was added to the bibliometric toolbox.

In the year 1998 Kopcsa, A. and Schiebel, E. (1998) published a mass point spring model to calculate the position of words in a two dimensional space. Nowadays, spring models and multidimensional scaling are models to map publications in a two or three-dimensional space. The advantages are the visualization, the representation of a huge number of bibliometric objects like documents, references, authors, keywords, etc. in two dimensions, and the visibility of relative positions of agglomerations of similar objects in the content based relation to each other. The processing time for algorithms for the positioning of objects in a two dimensional space grows with the square of the number of publications and limits the possible number of objects. Additionally, agglomerations could not be clearly delineated due to weak links between coupling documents. Schiebel, E. (2012) shows that many publications are unique and have little similarity to other documents by their reference vectors. They show a peripheral status and are to be qualified as in the similarity matrices and thus, also in the maps. Based on the sum similarity approach of Ahlgren, P. and Collander, C. (2009), the overall number of publications can be reduced without losing relevant documents. We introduce a threshold for the overall similarity of a document to better identify research fronts in a map of relative strong bibliographic coupled publications. The following procedure covers the assignment of publications to an agglomeration of similar documents based on the core document approach. The most relevant keywords and the titles describe the content of the identified research.

2 The data set and its treatment

The data source for the set of publications to explore the research issues for security research was the Web of Science (WoS) Database. We performed a topic search with the term "security" against the WoS Core collection for the year 2010. "Security" returned 11,318 entries for the year 2010.

The downloaded data set was further treated with the software BibTechMon. The data processing included the following steps and tables: import of the following fields: WoS ID, title, authors, keywords, abstracts and cited references; extraction of the references of all publications; generation of a table of relations between the WoS ID and the references; table of all references; matrix of the co-occurrences of the common number of cited references for each publication; matrix of Jij, as well as a similarity matrix that served as the matrix of spring forces.

The Jaccardindex is defined by

$$Jij = \frac{aij}{ai + aj - aij}$$

Jij standardizes the measurement of the similarity of the publications by their reference vectors: i=document I, j=document j, aij, number of common references, ai, number of references of document i, and aj number of references of document j.

The definition set for spring forces is -1 for Jij=0, and greater than 0 and equal or smaller than 1 for Jij.

Thus, all references served as coupling elements that defined the similarity of publications and allowed the positioning of the publications in a two dimensional space. First, all bibliographically coupled publications were mapped without any threshold by a spring algorithm, providing the first picture, see Figure 1.

The map of figure 1 was calculated with a spring model taking Jij similarity values as elasticity values for a spring model to position all publications in a two dimensional space. Each dot represents a document and its position in relation to all other ones. Dots with a sum similarity higher than 1.19 are coloured, the others are white. Local agglomerations show a set of publications with a high similarity among each other. Documents that have a small or none similarity to other documents are pushed away from the agglomerations to the periphery and are marked with white colour. The coloured background of the map is a 2D surface map of the local density of the number of papers per area unit weighted by the sum similarity of each document, see Schiebel (2012). The islands are areas with a higher concentration of a set of documents with strong bibliographically coupled documents; blue areas are populated by none or just a few dots with small or none similarities.

Figure 1: Science map of security relevant research in the year 2010

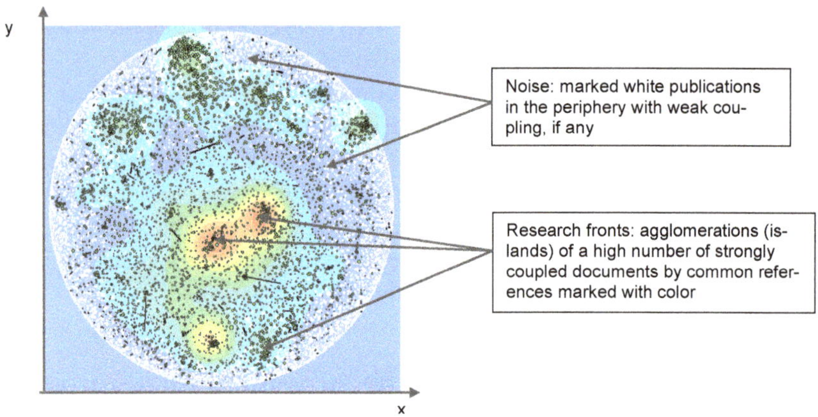

Documents mapped by a spring model, 10,956 publications with at least one reference, white dots are publications with a sum similarity Ji less than 1.19; the other 5,383 ones are coloured; background: spatial density of the number of publications weighted by the sum similarity, resolution 1/250, cosine smoothing with 50 dots; x, y local coordinates

In a next step, we discuss how dots with white colour that produce a noise in the map but take a remarkable part of calculating capacity can be excluded. We use the similarity of the documents to their coupling partners in the map. The sum Ji over j for one document represented the sum similarity for each document to all other documents:

$$Ji = \sum_j^n Jij,$$

with n=the number of all documents in the set.

Table 1: Jaccard matrix of similarities between exemplary publications – sum: sum similarity of each document, diagonal values: self-similarity

	doc1	**doc2**	**doc3**	**doc4**	**doc5**	**doc6**	**doc7**
doc1	1	0	0,05	0,01	0	0	0,02
doc2	0	1	0,33	0,36	0	0,51	0,73
doc3	0,05	0,33	1	0	0,05	0	0,32
doc4	0,01	0,36	0	1	0,32	0	0,56
doc5	0	0	0,05	0,32	1	0,05	0
doc6	0	0,51	0	0	0,05	1	0,14
doc7	0,02	0,73	0,32	0,56	0	0,14	1
sum	*1,1*	*2,9*	*1,8*	*2,3*	*1,4*	*1,7*	*2,8*

Table 1 shows the Jaccard matrix of 7 exemplary publications. Columns and rows are documents, and values are calculated by the Jaccard index of the co-occurrence of documents that is the common number of references of two publications. The last row is the sum of the column values. It is the so-called sum similarity. In Table 1, publication 2 and 7 have the highest sum similarity in the system. It means that they have a relative high number of common references with other documents and are part of a research work that is also published in other documents. The publications 1, 3, 5 and 6 have lower similarities. The documents of the second kind are not strongly or not at all linked to other ones, and thus, are not part of a collective and agreed effort in a research issue. The idea is to select only publications with a high sum similarity. First, we calculate the Distribution of all 11,318 publications that is shown in the next figure. Figure 2 shows the cumulative distribution function of the sum similarity of the Jaccard values. The distribution shows a dominating number of documents with a small similarity and just about 10 % the documents have a sum of Jaccard index higher than 3. The quick growing part of the distribution should include documents with a high similarity to documents with a similar knowledgebase represented by the references. Such documents are also called core documents; they are representatives of 'hot' issues in the research field. The smallest value of the sum similarity is 1 because the values of the diagonal of the Jaccard matrix – the self-similarities are 1.

Figure 2: Distribution function for the sum similarity (y: sum of Jaccard indexes) per publication; 10,956 documents with at least one reference; median = 1.19; x-axis: share of publications with sum similarity smaller than y

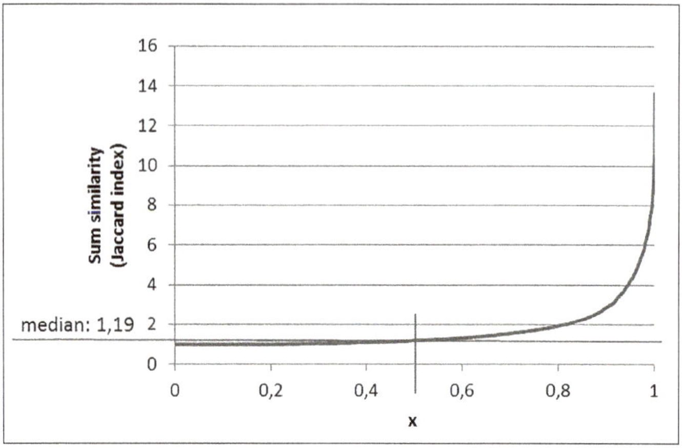

We take all documents with a higher similarity than the expected value. The median of the distribution represents the expected value.

The last step to identify research fronts is to recalculate the map with the same initial conditions as in Figure 1 but only with the coloured documents that contribute to agglomerations with a similarity higher than the threshold of 1.19 for the sum similarity. The results are shown in Figure 3.

Figure 3: Science map of research fronts in security relevant research, spring model, threshold of 1.19 for selected documents, number of publications 5383, left picture: surface map; right picture 3D-surface map – third dimension: local density of publications, digital resolution 1/250, cosine-smoothing with 50 dots

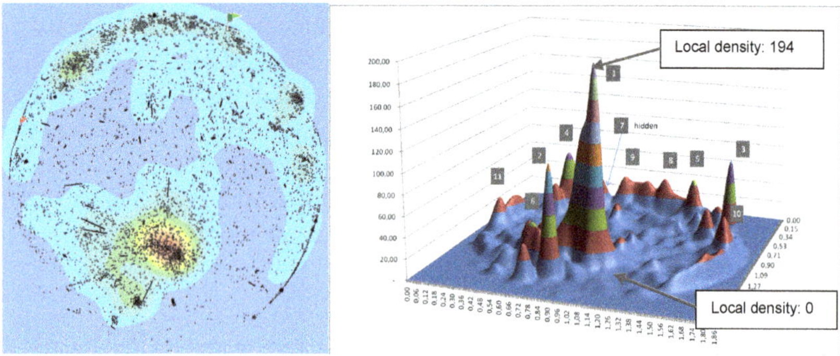

The map of Figure 3 represents only documents with a high similarity to their environment. The white dots that represent noisy publications are not included in the calculation. The time for one iteration was 10 seconds with 10956 publications with at least one reference and 2.3 seconds with the reduced set of 5383 documents. The calculation time was reduced by 23 %. The research fronts show a remarkable better structure in the 3D surface map of Figure 3 than in Figure 1. The third dimension z is defined by the local number of publications weighted by the local similarity. The weighted distribution shows areas of a high number of core documents that have a high number of strong links to documents in the same area, see Glänzel and Czerwon (1995). In the sense of a complex system, we can say that the authors of such papers are autonomous actors in a network of co-authorships and co-citations including previous work of colleagues and of their own and thus, are responsible for emerging research issues that become visible by such red zones.

3 Identification of research fronts

The calculation and visualisation of the map of bibliographically coupled documents enable the identification of hot areas of security relevant research by indicating hot zones. We need an additional procedure to decide which documents are part of the hot zone and what the content of the identified research issues is. The procedure is listed in Table 2:

Table 2: Procedure to assign documents to research fronts

Steps to assign documents to research fronts
Graphically assisted selection of documents in the center of a red heat zone
Selection of additional elements of the community (in the sense of a complex network community structure) by thresholds of similarity measured by more than two common references to define the extended community of similar documents
Visualization of the coupling elements (cited references) to examine the compactness of the selected community
Providing lists of tfidf ranked keywords, authors, organisations, cited references and abstracts of the publications of the extended community to name the research fronts

In Table 3, the 11 research fronts with the corresponding ID to Figure 3 are listed. It summarizes the research fronts and 4 indicators: the number of publications, the maximum local density, the number of links and the sum similarity of the documents that form a peak. The research fronts were named by their most relevant terms, measured by the modified tfidf that was introduced by Schiebel E. and Hörlesberger M. (2007), and additionally by the titles and abstracts of the documents.

Cryptography and quantum cryptography have the highest number of links and the highest similarity within its community. It reflects the high consistency in the research field with high research activity measured by the number of publications, the highest spatial density that is a result of publications with many links with high similarity.

Table 3: Research Fronts in security research

ID	Name	PUB	LOCDEN	NOLINK	LOCSIM
1	cryptography	412	194	9.060	627
2	quantum cryptography	99	106	4.741	393
3	psychology: attachment security	105	82	311	19
4	securities: investments, price models	155	67	754	50
5	food insecurity	60	45	281	19
6	image encryption	71	44	1.206	87
7	key management in wireless sensor networks	42	42	593	37
8	food availability	43	39	93	7
9	international security	54	38	111	8
10	power system security	108	34	417	36
11	THz spectrum dynamics analysis	53	34	178	27

PUB – number of publications
LOCDEN – maximum local density value (grid: 1/250; smoothing factor 20 for cosine smoothing)
NOLINK – number of links of all local publications
LOCSIM – sum of similarity values of all local publications

The researchers in this field comply with the research work done before and cite the publications of their colleagues; besides, there seems to be a considerable agreement on the common intellectual knowledge base that is represented by the commonly cited literature. Obviously, there might be some differences in the citation behavior in different disciplines. The third topic: *psychology: attachment security* is the social science that shows a complete different picture of the number of cited references. With nearly the same number of publications as quantum cryptography, we find only 311 links and also a much smaller similarity.

In the following chapter, we present a selection of two identified research fronts in security research.

3.1 Quantum cryptography

Quantum cryptography is a research field that applies quantum mechanical effects for the encryption, the decryption of cryptographic systems (Wikipedia 2012). The advantage of quantum cryptography lies in the fact that it allows the completion of various cryptographic tasks that are proven or conjectured to be impossible using only classical communication. In particular, quantum mechanics guarantees that measuring quantum data disturbs that data. This property can be used to detect an adversary's interference with a message.

The actual research in quantum cryptography is summarized in Table 4. The most relevant authors' keywords are: quantum key distribution (86.0); quantum cryptography (73.1); quantum secure direct communication (55.7); entanglement swapping (30.8); quantum secret sharing (27.4); deterministic secure quantum communication (23.3); single photons (21.7); cryptanalysis (18.4); decoy-state quantum key distribution (18.3); quantum dialogue (18.3). The keywords are sorted by descending tfidf that indicates the relevance of a keyword in a subset in relation to the occurrence in the whole set of documents.

Table 4: Research Front "quantum cryptography" in security research, keywords (with tfidf in brackets) and publications (with sum of Jaccard Index (sumsim) and number of links NoL))

List of keywords (tfidf value): quantum cryptography:
quantum key distribution (86.0); quantum cryptography (73.1); quantum secure direct communication (55.7); entanglement swapping (30.8); quantum secret sharing (27.4); deterministic secure quantum communication (23.3); single photons (21.7); cryptanalysis (18.4); decoy-state quantum key distribution (18.3); quantum dialogue (18.3)

WoS ID	**sumsim**	**NoL**	**Title**
WOS:000284959300002	7.58	78	Quantum secure dialogue based on single photons and controlled-not operations
WOS:000282920800004	7.55	72	SECURITY OF PRACTICAL PHASE-CODING QUANTUM KEY DISTRIBUTION

WoS ID	sumsim	NoL	Title
WOS:000284774100004	7.53	71	Experimental demonstration of phase-remapping attack in a practical quantum key distribution system
WOS:000277020600002	7.44	79	Continuous operation of high bit rate quantum key distribution
WOS:000275917600061	7.34	75	Quantum secure dialogue by using single photons

A possible commercial application that could utilize quantum cryptography is "two-party secure computation", in which two parties compare results of a computation without revealing the data used by each party to complete the computation. Since the two parties could be sitting at the same table, distance is not a problem.

There is also some work showing how to break conventional quantum cryptography systems based on various practical weaknesses in the way they are set up. Known loopholes can be closed, but new ones might appear.

Today, some commercial systems are available, mainly aimed at governments and corporations with high security requirements. A typical use is quantum key distribution. Whether quantum cryptography will replace classical cryptography techniques will depend on many factors, including transmission distance, expense, demonstrated threat of existing key exchange protocols, and ease of use. Optic fiber networks may offer the infrastructure for a more widespread use.

3.2 Food availability

The second research front that we present concerns the availability of food and the security related issues. Gregory and Brklacich (2005) examined the impact of climate change on food security. They report that: "*Dynamic interactions between and within the biogeophysical and human environments lead to the production, processing, distribution, preparation and consumption of food, resulting in food systems that underpin food security. Food security refers to the availability of food and one's access to it. A household is considered food-secure when its occupants do not live in hunger or fear of starvation. Food systems encompass food availability (production, distribution and exchange), food access (affordability, allocation and preference) and food utilization (nutritional and societal values and safety), so that food security is, therefore diminished when food systems are stressed. Such stresses may be induced by a range of factors in addition to climate change and/or other agents of environmental change and may be particularly severe when these factors act in combination.*

They summarize very clearly the problems and research issues on food security and availability. It becomes very actual in the ongoing refugee crisis, when people leave their country due to several reasons, and after all, due to food-insecure households.

Table 5. Research Front "food availability" in security research, keywords (with tfidf in brackets) and publications (with sum of Jaccard Index (sumsim) and number of links NoL))

List of keywords (tfidf): food availability

Adaptation (41.4); climate change (41.0); Green water (23.3); Irrigation (23.2); Virtual water (20.0); Food security (19.8); Vulnerability (19.3); phoma stem canker (18.3); Blue water (18.3); agriculture (17.2)

WoS ID	sumsim	NoL	Title
WOS:000287099300040	0.39	6	A Method for Spatially Explicit Assessment of Potential Risks of Food Insecurity
WOS:000272949100026	0.33	2	Medium- to Long-Run Implications of High Food Prices for Global Nutrition
WOS:000276097900011	0.3	5	Robust negative impacts of climate change on African agriculture
WOS:000284056900005	0.29	3	Adaptation to increasing severity of phoma stem canker on winter oilseed rape in the UK under climate change

The most relevant keywords are: Adaptation (41.4); climate change (41.0); Green water (23.3); Irrigation (23.2); Virtual water (20.0); Food security (19.8); Vulnerability (19.3); phoma stem canker (18.3); Blue water (18.3); agriculture (17.2). They indicate that environmental problems and climate change are main factors for food insecurity. Research work covers spatial assessment of potential risks of food insecurity, implications of high food prices for global nutrition, negative impacts of climate change on agriculture in Africa, and the particular case of the influence of climate change on increasing severity of phoma stem canker on winter oilseed rape in the UK, just to cite some core papers of published research work on food security.

4 Discussion and conclusions

The identification of research issues, especially emerging research fronts, with bibliometric methods is of growing interest for the evaluation and analysis of research agendas. The results are of value for the strategic positioning of expert organisations like universities, research organisations, and also for research and technology policy for regions, countries of for the European commission and the evaluation of science and technology policy funding frameworks. The identification of globally emerging research fronts is a requirement for the positioning of entities in the international science system. Additionally, science mapping is an important input for the understanding of science as a complex system. In our case, we focus on the identification of emerging issues.

Current bibliometric research on the delineation and identification of emerging research fronts shows much effort in this research agenda. Still, there are some limitations in relational bibliometrics. Very often, science maps and cluster analysis of bibliographic coupled publications accumulate publications that are not always consistent in their content or deliver suboptimal results for the delineation of (sub-) fields. Additionally, the computing time for relational problems grows at least with the square number of objects. We used the core document approach with a twofold effect: firstly, the reduction of the number of object extracting core documents that are of central meaning to current research, and secondly, the identification of additional documents with a very similar content to the core documents. That way we were able to collect sets of documents that indicate an emerging research issue.

Core documents have been suggested to be representatives of clusters that contain a huge amount of publications. In this work, we tested the other way round and used the core document idea to reduce the number of publications in a corpus. The results have demonstrated that the number of objects to be calculated in a computationally intensive spring algorithm is effective and can be controlled by the number of links with higher or lower similarity.

The identification of security relevant research fronts revealed topics from different disciplines like cryptography from physics, attachment theory from psychology or food security from social and political science.

References

ETCETERA (2013). ETECETERA Evaluation of Critical and Emerging Technologies for the Elaboration of a Security Research Agenda. http://www.etcetera-project.eu/.

Ahlgren, P. and Colliander, C. (2009). Document-document similarity approaches and science mapping: experimental comparison of five approaches. Journal of Informetrics, 3(1), 49–63.

Boyack, K. and Klavans, R. (2010). Co-Citation Analysis, Bibliographic Coupling, and Direct Citation: Which Citation Approach represents the Research Front Most Accurately?, JASIST 61(12): 2389–2404.

Chen, C. and Morris, S. (2003). Visualizing evolving networks: Minimum spanning trees versus pathfinder networks. Proceedings of IEEE Symposium on Information Visualization (pp 67–74), Seattle, WA: IEEE Computer Society Press.

European Commission (2007). Stepping up the fight against terrorism, Communication From The Commission To The European Parliament And The Council, Com(2007) 649 final Brussels, 6.11.2007.

Glänzel, W. and Czerwon, H.J. (1995). A new methodological approach to bibliographic coupling and its application to research-front and other core documents, Proceedings of 5th International Conference on Scientometrics and Informetrics, held in River Forest, IL, June 7–10 (1995), 167–176.

Glänzel, W. and Thijs, B. (2012). Using 'core documents' for the representation of clusters and topics. Scientometrics 88(1) 297–309. DOI: 10.1007/s11192-011-0347-4.

Gregory, P.J., Ingram, J.S.I. and Brklacich, M. (2005). Climate Change and Food Security. Phil. Trans. R. Soc. B 360, 2139–2148.

Janssens, F., Glänzel, W. & Moor, B. (2008). A hybrid mapping of information science. Scientometrics, 75(3), 607–631. DOI:10.1007/s11192-007-2002-7.

Kessler, M. M. (1963). Bibliographic coupling between scientific papers. American Documentation, 14(1), 10–25.

Kopcsa, A., Schiebel, E. (1998). Science and Technology Mapping: A New Iteration Model for Representing Multidimensional Relationships. Journal of the American Society for Information Science JASIS (1998), 49, 1, 7–17.

Price, D.D. (1965). Networks of scientific papers. Science, 149, 510–515.

Schiebel, E. and Hörlesberger, M. (2007). About the Identification of Technology Specific Keywords in Emerging Technologies: The Case of "Magnetoelectronics". Torres-Salinas, D., Moed, H. F. (eds.), Proceedings of ISSI 2007, 11th International Conference of the International Society for Scientometrics and Informetrics, Madrid, June 25th-27th, 2007, 691–69.

Schiebel, E. (2012). Visualization of Research Fronts and Knowledge Bases by Three-Dimensional Areal Densities of Bibliographically Coupled Publications and Co-Citations. Special Issue Scientometrics, DOI: 10.1007/s11192-012-0626-8, in press.

Shibata et al., (2009). Comparative Study on Methods of Detecting Research Fronts Using different Types of Citation. JASIST 60(3): 571–580.

Van Eck, N.J. and Waltman, L. (2010). Software survey: VOSviewer, a computer program for bibliometric mapping. Scientometrics, 84(2), 523–538.

Wikipedia (2012). Quantum cryptography from Wikipedia, the free encyclopedia, http://en.wikipedia.org/wiki/Quantum_cryptography. 19.3.2012. 16:32.

Yau, C., K., Porter, A., Newman, N., Suominen, A. (2014). Clustering scientific documents with topic modelling, Scientometrics 100: 767. DOI: 10.1007/s11192-014–1321-8.

Interventions in complex innovation systems

Michael Dinges, Karl-Heinz Leitner

The Austrian Competence Center Programmes: Past, present and future?

Abstract: *The Competence Center Programme was a major structural policy initiative since the 1990ies in Austria to support the collaboration between science and industry. The chapter investigates the strengths, weaknesses and characteristics of this specific form of co-operation in autonomous joint centers involving universities, research organisations and firms.*

1 Introduction

The competence center programmes (K programmes) have been one of the most important and most prominent RTI policy instruments in Austria since the 1990s. They constituted the first significant and highly funded programmes for the structural promotion of industry-science co-operations in Austria. The driving force behind the development of competence centers was the fact that cooperation between industry and science in Austria was found in the middle of the 1990s to be distinctly weak (cf. Pichler et al. 2007); the university sector was characterised by very low levels of R&D financing from the corporate sector (1.4 % of total university R&D spending) and a low degree of internationalisation (cf. Hutschenreiter et al. 1998). The OECD (2004), for example, stated that: *"Industry-Science relations (ISR) have been identified as one of the major weaknesses of Austria's innovation system. In fact, ISR-related indicators show below-average values in international comparisons."*

The government responded with its K programmes to international developments in the direction of greater promotion of the cooperation between industry and science, which were advanced as a key instrument for accelerating the transfer of knowledge and competition. There were, however, some specific Austrian challenges to be addressed. Firstly, there was a need to change the strategy of Austrian corporate research, in order to rely more on collaborations with scientific institutions for gaining access to new knowledge and basic scientific results. Secondly, there was the matter of also renewing the Austrian industrial structure by means of promoting scientifically based technology fields (such as ICT, biotechnology, materials engineering), where scientific research played a key role in technological change. In addition, it was necessary to effect a certain reorientation of the public research institutions themselves. The co-operation with companies should also be regarded to an increasing degree by universities as a strategic resource. Another challenge in this respect was the fact that incentives at the universities for cooperative research with companies were low until the implementation of UG (Universities Act) 2002.

The idea of giving competence centers a key role in Austrian research and technology policy was put forward in a strategy paper for technology policy in the year 1997 (Schmidt and Hochleitner 1997). A working group of the former BMWV (Federal Ministry for Science and Transport) subsequently advanced the development of the competence center programme Kplus (cf. Biegelbauer 2006, 2007). The strategic orientation of the competence center programme had already been clearly apparent in the policy paper ("K+ Research Competence plus Economic Competence", BMWV 1997). In 1998 the competence center programme Kplus, whose needs structure was more "knowledge driven" (cf. Edler et al. 2004) was launched by the BMWV (now BMVIT (Federal Ministry for Transport Innovation and Technology)). One year later, the former BMWA (Federal Ministry for Economic Affairs (now BMWFW (Federal Ministry of Science, Research and Economy)) launched its own "innovation-driven" programme (cf. Edler et al. 2004), which, compared to Kplus, was closer to the market.

Throughout the entire duration of the competence center programme from 2001 to 2009, the accumulated public expenditure for the Kplus programme amounted to € 432 million and € 399 for Kind/Knet. This adds up to a total expenditure of € 831 million throughout the entire duration of the programme (cf. Schibany et al. 2013).

In the year 2006, the competence center programmes Kplus and K_Ind/K_Net were bundled together to form a new competence center programme COMET, which is implemented via three programme lines: K1 centers comprising rather smaller dimensioned competence centers, whose orientation corresponds more or less to that of the older Kplus centers; K2 centers, which are considerably larger than the K1 centers and their interest is focused more on R&D excellence and internationalisation; and K projects, which are industry-science networking projects that have to comprise at least three company partners and one scientific partner.

The topics for all K center and K project calls are open. The main distinguishing features of the three COMET programme lines concern the demands placed on the funded institutions with regard to internationality, project volume, duration and (physical) infrastructure.

The first COMET centers started their operative activities in 2008. In the course of the calling process carried out in the COMET programme all 17 centers from the Kplus programme (in part as reconfigurations) received further funding. Only two centers emerged in the K projects – smallest funding instrument. Five of the 28 funded K_ind/K_net centers were discontinued. Ten K_Ind/K_Net centers emerged merely as K projects. The fusing of complementary K centers resulted in the larger consortia proposed in the 2004 assessment (Edler et al. 2004) during the period between 2008 and 2010.

So, in the meantime, the Austrian competence center programmes can look back on a history stretching over around 20 years. In particular, the programme design and the implementation of Kplus played a decisive role in the competence center programmes being classified internationally as "good practice" (cf. OECD 2004, Janger and Friesenbichler 2008).

Several surveys and evaluations have examined various aspects since the foundation of the programme (Schibany et al. 2013, Edler et al. 2004, Biegelbauer 2006). Based on these and other findings, the competence centers will be analysed in the following in respect of their core tasks, whereby our main focus will encompass the questions of joint knowledge production, knowledge transfer and innovation performance.

2 Characteristics of the Competence Centers

The following will look at the key characteristics of the K programmes (Table 1 gives an overview):

Table 1: Key characteristics of the Competence Center Programmes in Austria

			COMET (since 2007)		
	Kplus (1998)	**K_Ind/Net**	**K-Projects**	**K1-Centers**	**K2-Centers**
Duration:	7 years (4+3)	4 years	3–5 years	8 years (4+4)	10 years (5+5)
Public Funding:	Max. 60 %	Max. 60 %	40–50 % (incl. min. 5 % SP)	40–55 % (incl. min. 5 % SP)	40–55 % (incl. min. 5 % SP)
Public Funding / year:	2–4 million p.a.	Ca. € 0.5–4 million p.a. (K_Ind)	max. € 0.45 million p.a.	max. € 1.7 million p.a.	max. € 5 million p.a.
Number of Centers/ Networks	17	16 Centers 12 Networks	Ca.35	16	5
Key Objectives	– Improved exploitation of existing knowledge and production of new knowledge by means of long-term research programmes – Location attractiveness and ensuring Austria as a research location – Achieve national competence and critical mass in industry-oriented research – Boost the Austrian chances of participation in international research programmes	– Exploitation and bundling or focussing of the existing knowledge in the companies and public research institutions for innovation – Stimulation and sustainable increase of private R&D expenditure – Bundling of resources for the promotion of technology clusters – Development of expertise to improve the chances of participation in international funding programmes – Spin-offs in the research and technology-related sector	– Development of new expertise by initiating and supporting strategically aligned and long-term research co-operations between industry and science at the highest level – To develop and ensure technical leadership for Austrian companies – To further develop and pool existing strengths and embrace international research expertise in order to consolidate Austria's position as a research location in the long term		
Funding criteria	– Research programme defined jointly by science and industry – Research competence of key researchers – Human resources development – Better access to international R&I funding	– Research and innovation demand of enterprises – Scientific qualification (also enterprises) – Application opportunities Business Plan	– Research programme defined jointly by science and industry – Research competence and connection to science – Relevance of implementation in industry – Quality of collaboration between science and industry		

Source: Own compilation based upon: Edler et al. (2004), Schibany et al. (2013), Dinges et al. (2015)

2.1 Technological Focus

17 Kplus competence centers and 28 K_Ind/K_Net centers or networks were funded in the first programme phase of the competence center programmes (Kplus and K_Ind/K_Net). The centers and networks were spread across all the Austrian Federal Provinces, whereby, in view of both location and the number of partners, particular importance was placed on the strongest industrial provinces Upper Austria and Styria followed by Vienna, Lower Austria and Tyrol. The main thematic focus was on the fields of information technology, electronics, mechatronics and sensor technology followed by materials engineering and life sciences. The centers also addressed other thematic fields such as automotive engineering, process technology, energy technology, the environment and wood. This fundamental orientation did not change in the second phase of the competence center programme (COMET). Key priorities of the K1 & K2 centers included the research sectors mechanical engineering/instrument making (38 %), chemical science (18 %), electrical engineering /electronics (13 %), information technology (9 %), mathematics/informatics (6 %) and medical chemistry, med. physics and physiology (5 %)[1].

So for the main part the K programmes addressed and still address traditionally strong fields of Austrian corporate R&D (e.g. material sciences), although a high percentage of enterprises involved (compared to the R&D survey) can be found in the instrumentation equipment, electronic component, medical devices and ICT service branches. Combinations of research topics were limited to only a few centers and topics in the first phase of the programme. Interdisciplinary active centers were rather an exception. Nothing has changed fundamentally in this respect with regard to the COMET centers.

2.2 Partner structures

The number and relative importance of actors from different contexts (universities, enterprises) is a key characteristic of the K programmes. In the first phase of the programmes, there were in total 710 (28 %) company participations and 213 (22 %) participations from the scientific field, whereby the Kplus programme attracted considerably more company partners (528) than the K_ind/K_net programme (182). Altogether there were 657 companies behind the total number of 710 company partnerships, whereby multiple participation in both programmes was restricted mainly to large companies. In the second programme phase

1 https://science.apa.at/docroot/apa-science/images/factbox_dossier_comet_1200.png

(COMET) it was possible to increase the partner participations to 1,235 (+35 %) (incl. multiple responses) (Last updated 2015[2], FFG Comet Monitoring Report 2015). Around two thirds of the partners were companies and one third were scientific institutions, indicating that it had also been possible to increase the percentage of scientific partners in comparison to the predecessor programmes.

The span of partner participations fluctuated significantly in both the first and second programme phases. There were between 13 to 41 partners in Kplus and 2 to 22 partners in K_ind/K_net. The minimum number of participations in the successor programme COMET in the year 2015 stood at 18 in the K1 centers and 52 in the K2 centers, while the maximum values were 123 and 140 in the K1 and K2 centers, respectively.

Even the K programmes in the first phase addressed in particular large (for Austria) companies in comparison to the total structure of companies that undertook R&D. With regard to the size of the companies, 40 % of the companies that participated in the K programmes were large companies with more than 250 employees and therefore, according to the 2009 R&D survey, far above the Austrian average (14 %). This situation did not change in the second programme phase, where large companies (in this case, with more than 500 employees) made up 40 %, 25 % and 30 % of all company participations in the K2 and K1 centers and K projects, respectively.

2.3 International profiling

The competence centers were also used in varying degrees as an instrument of international profiling and helped establish co-operation relationships on an international level. There were 150 participations of foreign companies in the first programme phase, all of which concerned the Kplus programme. Kind/net did not realise any noteworthy international partnerships. According to the evaluation report from 2013 (Schibany et al. 2013), the Kplus centers had an average of 8.7 foreign company participations and therefore a 25 % share of foreign companies in all company participations. The co-operation relationships with the foreign company partners were, however, extremely heterogeneous between the centers. It was possible to increase the participation of foreign company partners in COMET to 282 participations (cf. FFG Impact Monitoring Report 2015), although these remained equally heterogeneous. Germany is still the main partner country with regard to international activities (around 50 % in COMET vs. 80 % in Kplus),

2 https://www.ffg.at/sites/default/files/allgemeine_downloads/strukturprogramme/monitoringbericht_2014_15_k1k2_zentren.pdf

although meanwhile the neighbouring countries Italy and Switzerland also play a certain role in the programme. In total, 70 % of all international participations in COMET involved these countries.

2.4 Regional ties

With respect to the regional distribution of partners, a high regional importance can be determined for the K centers. The average percentage of company partners from the core region, for example, was 46 % in the Kplus programme. In COMET, too, on average, 44 % of the national company partners in K2 centers came from the Federal Province in which the center had its head office (minimum 25 %, maximum 70 %), this value was even 50 % in the case of K1-type centers (minimum 12 %, maximum 81 %). With very few exceptions, it can therefore be claimed that regional ties to the centers can be characterized as very high.

The financing structure of K programmes that define a non-K area and in which the centers should attract research contracts underlines the importance of the regional ties to the centers. The so-called non-K area was introduced in the Kplus programme in order to guarantee the long-term capacity to act of a research unit organized as a GmbH and to reduce the share of public funding over time. Such non-K areas are intended to enable the centers to acquire R&D and service contracts over and above their research activities within the framework of Kplus.

According to the evaluation report, financing by the non-K area in the year 2013 was, on average, 27 % of the total budget, although this figure varied significantly between the K centers. However, the non-K area never achieved a magnitude in any of the centers that could have ensured a long-term existence in the current form without public funding.

Four basic types of center can be discerned based on these characteristics and the analyses carried out within the framework of the ex-post evaluation (Schibany et al. 2013):

1. Regional platforms that assume an important function of the regional economy. They often have a more practical orientation, build upon and strengthen existing co-operations. As they are generally geared towards the needs of the regional economy, they function more as platforms than as strong, independent actors.
2. New networks that bundle the expertise in several Federal Provinces while breaking new ground with co-operations between science and industry. Their activities are geared primarily towards the exploitation of industrial partners, but can also involve establishing scientific expertise and a corresponding profile development.

3. Lighthouses that generally build upon well-established co-operations between science and industry and so, reflect the well-known strengths of the Austrian industry and/or research. The focus here is on clearly visible scientific profiling and the creation of a new quality of co-operations. As strong actors they seek to establish access to international top research and co-operations.
4. Centers of the future that seek to establish and bundle expertise in relatively new and less well-established fields and aim at developing to become a corporate actor. Several actors act in concert here to exploit synergy potential jointly and to achieve international/national visibility.

3 Strengths, weaknesses and momentum

The key findings of the most extensive evaluation of the Kplus programmes to date indicate both strengths and weaknesses in the K programmes.

Overall, a strengthening of the co-operation between the companies and research institutions involved was determined. Center-wide transversal projects and joint research planning by all organisations involved were of key importance in this respect. At the same time, a growing tendency has been observed in some centers towards independent, non-cooperative activities in individual organisations.

3.1 Scientific output

There is a great difference with regard to the results achieved in the centers. The scientific output outweighed in some and the economic output in others (e.g. in the form of patent applications). The overall output also fluctuated significantly, so that not all centers were deemed to have fulfilled the demands on excellence. As a general rule, the orientation of centers depends to a large extent on key researchers. A stronger relative importance of public research institutions in centers correlates positively with the importance of key researchers. It can therefore be concluded that the research of the centers where universities assume this function – and open up opportunities – is closer to the market and more problem orientated.

For the first programme phase, the evaluation performed was only possible to consider the scientific output of the Kplus programme, due to the limited data available (see Schibany et al. 2013). The pictures revealed by the evaluation were very heterogenic. In particular, the fact that in several centers hardly any dissertations were completed, only a few publications achieved and no patents applications submitted can only be assessed as unsatisfactory, albeit center-specific, for a strategically orientated research programme such as Kplus. However, it must be taken into consideration, especially where patents are concerned, that the allocation of achieved patents from K activities to the K centers was incomplete and difficult to determine.

The 2015 COMET impact analysis indicates that the attempt to promote excellence via the K2 centers has been successful only in part and heterogeneity remains high (Dinges et al. 2015). There are only relative minor differences between the K1 and K2 programme lines with regard to the scientific and technological output. It is surprising that the K2 programme line, with its considerably higher level of staffing and larger annual budget, was unable to achieve a higher output per full-time equivalent (FTE) than the K1 programme line and was even below the level of the K1 programme line in respect of realised patents and scientific publications per million Euro of funding. Apart from the high diversification in the publication output between the centers, it was also observed that in particular, the level of publications in the centers depended more on the number of employed PhD students than on the number of employed scientists.

3.2 Impact on firm innovation

According to the ex-post evaluation of the first programme phase (Kplus/K_Ind_Knet), the impacts on companies were modest. Astonishingly, potential additionality effects seemed to be limited due to an already existing high level of intersectoral co-operation before participation to the programme. A large number of companies claimed that they had already carried out comparable research projects (or co-operations) before the existence of K centers. However, the K programmes contributed to making the form of co-operation within the framework of the K centers more intensive, innovative and institutionalised.

The evaluation of the importance of the research projects for companies indicated that for a large percentage of the funded companies these were only pertinent for peripheral areas of the company's own business or technology strategy. Orientation towards R&D projects with higher risk exposure seem to have taken place to only a limited extent. The evaluation as to whether the co-operation in a competence center in fact brought forth any significant product or process innovations indicated that between 40 and 50 % of the companies could not report any specific innovation outputs.

It would appear that any positive overspill of knowledge deriving from internationalisation impacts as a result of the participation of foreign companies in the Kplus centers was also low. Here, too, the impression is gained that many of the foreign companies had already been co-operating with Austrian companies and scientific institutions in Austria before the start of the funding scheme.

Overall, based on the data available, the 2013 evaluation yielded a rather sober impression of the impacts of the programme on the companies involved. In particular, when those companies were funded, they already exhibited a high level of R&D activities and had already gained experience in co-operations with research institutions. These could, of course, have been further intensified by participation in the programme. However, the quality of the research projects (with regard to co-operation partners, duration of the research projects and intensity of the co-operation) was probably not fundamentally any different from those research projects that the funded companies carried out even without the existence of the competence centers. The diverging results indicate that radical innovations based on extremely risky R&D projects hardly ever took place or even not at all.

The 2015 impact analysis of the COMET programme phase (Dinges et al. 2015) indicates that the cooperative research in the K centers played a significant role in boosting the promotion of competence (80 % of the companies surveyed) and the innovative output of the companies participating in the K centers.

The research activities in COMET built primarily upon the long-term, well-established co-operation structures between the COMET centers, company partners and scientific institutions. Whereby the planning of the research activities is focussed clearly on specific, strongly application-orientated problems of the individual companies.

During the period 2010–2015, 65 % of the innovation active COMET companies managed to create product or service innovations induced by COMET. In some cases the range of innovations derived from COMET, in addition to incremental ones, also included fundamental new developments with a considerable impact on the implementing organisations. These include, for example:

- The development of new platforms that are now commercialised by companies in partnership with K centers.
- The development of new products that were only possible based on the entirely new technological developments of the K centers and partners. The consolidation of the technological leadership of Austrian companies on the world market and a contribution to securing the production location.
- The development of innovations that fundamentally accelerated production processes by means of a technological integration of sales and planning.
- The development and implementation of technologies that in individual cases resulted in the establishment of new companies.

Only a few of the companies surveyed noticed any impact in respect of higher resource efficiency in production, the gaining of new customer groups and markets or an increase in the number of employees in the business segment as a result of RTI activities in COMET. Neither did participation in COMET lead to an increase in the R&D budget in the company, according to the majority of the companies surveyed.

3.3 Multi-firm projects

All studies about the K programmes reveal that often too few multi-firm projects were implemented or established in all programmes or phases. The ideal type of research project in a competence center can be defined as an inter-disciplinary, multi-firm project in which the research is precompetitive in character. Such projects presumably exhibit the greatest spill-over effects. The incentive structure for the execution of such projects in the centers has proven complex and by no means entirely favourable for the realisation of such planned research. There is very little incentive to implement multi-firm projects in particular where the company partners are competitors in the same product markets. This resulted in many companies using their participation in the K programmes for the execution of bilateral projects.

For this reason, the design of research topics should be precompetitive and strategic and thus, enable several industrial partners to co-operate together in a multi-firm project. Despite the great diversity of project combinations in the centers, the focus has to a great extent become concentrated on bilateral projects. The COMET centers were unable to carry out co-operations between several companies to the desired degree due to the strong application-orientation of the research. The COMET impact analysis indicates that true multilateral projects involving several company partners were typically only possible under the following specific conditions:

- If co-operations between suppliers and producers are developed along the value chain: a typical example of this is the automotive industry, in which strong ties and development requirements exist between suppliers and car producers (OEMs). If the cooperative projects operate along the value-added chain, then, according to one surveyed company, processing via COMET offers the possibility of Austrian suppliers securing common exploitation rights with OEMs, which is often not possible in direct projects between suppliers and OEMs.
- If strong innovation pressure arising from regulations constitutes challenges for an entire branch, in particular stricter environmental standards, require-

ments for new developments in the field of electromobility etc. this can result in a pressure to innovate that necessitates joint developments within an entire branch.
- If the industrial structure is such that there is hardly any company with sufficient R&D capacity to carry out applied research (e.g. many companies in the bioenergy or food industries). The joint use of R&D infrastructures taking obligations of confidentiality into consideration impels companies to work cooperatively.
- If an area of technology, a process or method that can be used in different branches is processed by the COMET center/project (e.g. integrated process analytics in production processes).
- If the research and development work is technologically so far upstream that direct marketing is not possible in the foreseeable future. This was not normally the case in COMET.

The programmes tended to have a stronger impact where universities were concerned. Overall, the K programmes had a marked impact on the strategy and profiling at the technically orientated universities in the first programme phase. They were a driving force at some universities (Linz, Leoben, TU Graz) for specialisation and had a significant impact on their orientation.

4 Discussion and outlook

Establishing joint institutions between research institutes and firms is a specific form of knowledge transfer. The competence center programme played a key role in the promotion of this independent institutional model in Austria.[3]

This specific type of knowledge transfer and collaboration is faced with specific challenges. In general, industry and science pursue two different rationalities. According to Luhmann (1991), both industry and science display characteristics of autonomous social systems. A social system can be defined as autonomous if the persons and institutions involved pursue a rationality that is inherent to the system. The key elements of a system rationality are system-specific goals and behavioural patterns or rules that can make interaction between different systems difficult. Goals may be incompatible or conflicting, behavioural patterns may be unacceptable or incomprehensible for the other party.

3 In contrast, the focus of the Christian Doppler laboratories, in which the involvement of industry consists mainly in the formulation of a specific research need, is primarily on scientific research.

This lends particular relevance to the question as to how the interaction between the actors in different social systems can and should be organised so that actors from all the participating systems can benefit from the co-operation. Seen in the light of the importance of the system-specific rationalities of the two social systems "industry" and "science", the co-operation can, however, stimulate innovations effectively if the persons or organisations involved in the respective systems are able to pursue their prevailing system rationalities to an adequate degree (Kaufmann 2009).

The co-operation is only expedient for both parties if its results are regarded as worthwhile according to both system rationalities, in that added value vis-a-vis R&D activities is generated and remains within the respective limits of each system. For industry, such added value consists primarily in the fact that innovation activity is not only enhanced per se, but that it results in particular in more far-reaching, radical innovations. The additional benefits for science include the generation of new issues, specific information and practical data that would not otherwise be accessible and knowledge of the application context. There is also the chance of growth and attaining critical mass. A key criterion for the success of a center is to find ways of circumventing conflicts of goals. This involves formulating consistent goal hierarchies and success criteria and system-compatible incentive systems (Kaufmann 2014). If, for example, funds raised in the form of R&D contracts are the criteria for the evaluation of success, then scientific output in the form of publications will eventually decline.

Added value for both social systems must be generated from the co-operation in the long term for both the participants from academia and those from industry. The danger here is that, the work of scientific partners is too industry-orientated, so no emphasis is placed on possible potential for radical innovations.

Without doubt, the K centers have helped to enhance the co-operation structure between science and industry and to establish long-term institutional relationships. A glance at the statistics clearly indicates that the intensity of co-operations between companies and academia is also extremely well developed at an international level. Austria is ranked around third place within the European Union, concerning the co-operation of innovative companies and universities (Leitner et al. 2015).

The cooperation between science and industry is not the panacea for boosting the innovation capacity and competitiveness of the latter. There are signs of saturation in the funding of R&D co-operations.

The 2015 impact analysis of the second programme phase indicated that, in addition to very positive fundamental impacts, there are some basic questions with regard to the incentive structures and functions of K centers for the future, particularly in respect of the strategic role and relevance of COMET centers for scientific partners. In view of the intense changes taking place in the university environment, there is a growing tendency to regard the type of research funding for COMET centers as less incentive-compatible for universities. Points criticised in this respect include, among others, the fact that scientific partners are sometimes invited to take part in projects solely for the purpose of gaining specific expertise or to provide a specific research service. Furthermore, although the centers support the intensification of topics, they hardly ever act as direct catalysts from the application point of view (e.g. with regard to new course content or new methods). In view of the growing necessity for universities themselves to acquire third-party funding, application-orientated universities certainly find themselves in competition with institutionally powerful centers. And this is the case despite the fact that it has been proven that participation in COMET generates added value, even for scientific partners.

In the future, funding of the more organisationally costly and more strictly regulated organisation models such as COMET centers could be focussed on a few large and strategically important organisations. Funding of the co-operation between science and industry should focus primarily on intensification, openness and new pathways, not on further expansion. It would appear more expedient to provide more generous and long-term funding for a few joint R&D projects of particular strategic importance rather than support as many new partnerships as possible and accept the fact that many of these will contribute very little new impetus to research in science and innovation in industry.

References

Biegelbauer, P. (2006). Innovations in Innovation Policy Making: The Austrian Competence Centre Programme K+.HIS, Reihe Soziologie, May 2006. Wien.

Biegelbauer, P. (2007). Learning from abroad: the Austrian competence centre programme Kplus. Science and Public Policy, 34(9), S. 606–618.

BMWV (1997). Kplus – Forschungskompetenz plus Wirtschaftskompetenz: Vorhabensbericht des Bundesministeriums für Wissenschaft und Verkehr zur Errichtung von Kompetenzzentren in Österreich. Wien.

Dinges, M., Ploder, M., Wepner, B., Zahradnik, G. Streicher, J. Linshalm, E. (2015). Wirkungsanalyse 2015 des österreichischen Kompetenzzentrenprogramms COMET, AIT – Innovation Systems Report, AIT IS Report Vol. 111, Austrian Institute of Technology, Wien.

Edler, J., Lo, V. and Sheikh, S. (2004). Assessment "Zukunft der Kompetenzzentrenprogramme (Kplus und K_ind/K_net) und Zukunft der Kompetenzzentren". Endbericht. Studie im Auftrag von BMVIT und BMWA.

Hutschenreiter, G., Knoll, N., Paier, M., Ohler, F. (1998). Technologiebericht 1997, Studie des Österreichischen Instituts für Wirtschaftsforschung und des Österreichischen Forschungszentrums Seibersdorf, Mai 1998, Wien.

Janger, J. and Friesenbichler, K. (2008). Kplus Competence Centres (Austria). In: OECD, A Review of Local Economicand Employment Development Policy Approaches in OECD Countries, 192–203.

Kaufmann, A. (2009). Organisationsformen der Zusammenarbeit von Wissenschaft und Wirtschaft in Österreich, in: Leitner, K-H., Weber, M., Fröhlich, J. (2009) (eds.): Innovationsforschung und Technologiepolitik in Österreich: Neue Perspektiven und Gestaltungsmöglichkeiten, Studienverlag, Innsbruck.

Leitner, K-H., Dachs, B., Degelsegger, A. et al. (2015). Stärkefelder im Innovationssystem: Wissenschaftliche Profilbildung und wirtschaftliche Synergien, project commissioned by the Federal Ministry of Science, Research and Economy, Wien.

Luhmann, N. (1991). Soziale Systeme: Grundriß einer allgemeinen Theorie. Suhrkamp, Frankfurt am Main.

OECD (2004). Public-Private Partnerships for Research and Innovation: An Evaluation of the Austrian Experience. Paris.

Pichler, R., Stampfer, M. and Hofer, R. (2007). Forschung, Geld und Politik: Die staatliche Forschungsförderung in Österreich: 1945-2005. Studienverlag, Innsbruck.

Schibany, A., Dinges, M., Reiner, C., Reidl, S., Hofer, R., Marbler, F., Leitner, K-H., et al. (2013). Ex-post Evaluierung der Kompetenzzentrenprogramme Kplus und K_ind/K_net, Study commissioned by the Federal Federal Ministry for Transport, Innovation and Technology and the Federal Ministry for Economic Affairs, June, Wien.

Schmidt, A. and Hochleitner, A. (1997). Forschung und Wettbewerb. Technologieoffensive für das 21. Jahrhundert; Bericht an die Bundesregierung, Wien.

University Organisation and Studies Act (Universities Act 2002). University Organisation Amendment Act and Universities of the Arts Organisation Amendment Act, No. 120/2002 / 9th August.

Eva Buchinger

Innovation policy lab to design intervention in public procurement

Abstract: *Public procurement promoting innovation (PPPI) is high on the policy agenda. The aim is to provide better services for citizens and create attractive markets for the industry. Since procurement is, in its nature, conservative and risk-averse, Austria runs a PPPI initiative. A policy lab approach is used for the optimal design and implementation of this intervention.*

1 Introduction

Demand-side instruments such as public procurement, regulation and standards are nowadays an important issue in innovation policy. Together with supply-side instruments such as subsidies for research and development, they should form an effective innovation policy mix[1]. Public procurement of innovation is presently the most prominent demand-side instrument because of its huge economic relevance[2]. On the one hand, hopes are high that public procurement can provide a powerful lever for innovation. On the other hand, procurement is in its nature conservative and risk-averse, and the innovation stimulating purchasing power of public entities remains somewhat unexploited. Thus, intervention is required and justified from an innovation policy perspective.

Austria, which belongs meanwhile to the group of advanced countries in this field[3], has been using a policy lab approach for the optimal design and conduct of a policy intervention under the label "Public Procurement Promoting Innovation (PPPI)". This text deals with the question how far PPPI has been institutionalized up to now. Thereby, the PPPI Lab, its context, its rollout and its outcome will be described and the progress of the institutionalization of PPPI will be valued. Before that, the rationales of public procurement of innovation and the systems theoretical background will be outlined.

---

1 See e.g. EC 2007/C/799, 2007/C/860; OECD 2011, 2014; SBIR-STTR-IPC 2014; Edler and Georghiou 2007; Edquist et al. 2015.

2 Public procurement accounts for over 14 % of European Union gross domestic product (GDP) (EC 2016b).

3 See for overviews ERAC 2015; OECD 2016; Buchinger 2016.

Box 1: Definition of innovation procurement

Since public procurement is object of European law, the definition of innovation procurement used here follows the respective EU Directive, which has been dutifully translated into national Austrian law (OJEU 2014/L/94–24, 2014/L/94–25; BGBl 2006/17; BKA 2017): (i) Procurement is defined as the acquisition of the 'execution of work', the 'supply of products' or the 'provision of services' for consideration by means of a public contract (between a contracting authority as buyer and an economic operator as supplier). (ii) Innovation procurement is defined as the acquisition of a new or significantly improved product, service or process, including but not limited to production, building or construction processes, a new marketing method, or a new organisational method in business practices, workplace organisation or external relations. (iii) Research & development (R&D) services are covered by procurement law only if the benefits accrue exclusively to the contracting authority for its use in the conduct of its own affairs, and the service provided is wholly remunerated by the contracting authority. Otherwise, they are exempted and labelled pre-commercial public procurement of R&D (PCP) in distinction to (i–iii) commercial procurement of innovative solutions (PPI). (iv) The new procedure innovation partnership combines PCP and PPI.

2 Rationale: Why public procurement promoting innovation (PPPI)?

Boosting innovation through procurement is not new. It has been done occasionally since decades to modernize public infrastructures. Examples are implementing e-health devices, demanding renewable energies, introducing intelligent mobility solutions, and modernizing safety & security facilities. New is the aim to achieve higher rates of innovation procurement in all areas. This aim follows a rationale consisting of three strands of considerations.

First, enhance public administration's effectiveness. Public entities do not exist for their own sake. Their mission is the delivery of high-quality services in areas such as health, economic & social inclusion, safety & security, and sustainability by simultaneously following the low-cost principle. To provide continuously qualitative output and throughput, they need also continuously (and not only occasionally) qualitative input. That is, qualitative output in the form of services for citizens (external performance); qualitative throughput in the form of efficient processes (internal performance); and qualitative input in the form of procurement of innovative works, supplies, and services[4]. Thereby public entities can be

4 Works means the outcome of building or civil engineering works taken as a whole, which is sufficient in itself to fulfil an economic or technical function. Supply means the purchase, lease, rental or hire-purchase, with or without an option to buy, of prod-

re-active and procure best available solutions already offered by the market. Or, they can be pro-active and demand innovation/s in their call for tenders. Both procurement strategies do have their advantages and disadvantages and must be carefully chosen.

Second, address societal challenges. The procurement of innovative works, supplies and services can address societal challenges beyond the delivery of low-cost and high-quality services for citizens. One of these challenges is already well established as green procurement. The 3rd edition of the Handbook on Green Public Procurement (EC 2016a), the Green Procurement Guidance (EPA 2014) and the collection Going Green (OECD 2015) with all their good practice examples provide evidence for the meanwhile widespread acceptance and steadily growing conduct of environmentally friendly public procurement. Another societal challenge begins to be recognized as social procurement. The guide Public Procurement for Social Progress (SocialPlatform 2015) shows how social considerations such as social fairness and inclusion[5] can be an issue throughout the entire procurement process. Furthermore, easier access of small and medium sized enterprises (SMEs) to public procurement contracts is a challenge. The rational is that higher SME participation results in increased competition and leads therefore to better value for money for public entities and allows SMEs to grow and create jobs and thus contribute to societal welfare (EC 2008). Addressing these challenges represents a change within procurement law from the principle of selecting the 'most economically advantageous' tender to the principle of selecting the tender with the 'best price-quality ratio'.

Third, mobilize and direct resources for innovation. More than a decade ago, several expert groups (EC-Expert-Group 2006, 2005, 2003) suggested a paradigm shift going well beyond the narrow domain of research & development and innovation policy. Among others, public entities should act as 'intelligent customers' and become lead users and even create lead markets to better direct research and innovation to end-needs. This was later on reflected in innovation policy strategies such as the European flagship initiative Innovation Union (EC 2010/C/546) and the OECD Innovation Strategy (OECD 2010). Especially in sectors such as energy, transport, waste management, social protection and

---

ucts. Service means provision of services other than those referred to in works. See the respective European procurement directives (OJEU 2014/L/94–24, 2014/L/94–25).

5 This means, among others, integration in society via employment and occupation, guaranteeing equal job-opportunities for all, fair payment (exclusion of wage dumping), sheltered workplaces, integration of disabled persons and disadvantaged and socially marginalized groups, and gender equality.

educational and health services, public entities are often principal buyers and therefore (potentially) very powerful in stimulating innovation.

These three strands of consideration converge in the one overall rationale: capitalize on innovative solutions to modernize public services for the benefit of citizens' immediate and long-term wellbeing.

3 Intervening system and addressed system in PPPI

On the macro societal level, the intervening system of the PPPI initiative is politics and the addressed system is economy. On the organisational level the addressees are economic operators – i.e. enterprises or other forms of suppliers[6] – which are expected to increase their innovation performance following the demand of public buyers. According to social systems theory, societal as well as organisational systems are complex 'operationally closed' entities (Box 2). Operationally closed systems cannot be controlled from outside, i.e. being the intervention 'object' of an outside located intervention 'subject' (Luhmann 1997). But they are open to various kinds of signals from their environment and can be brought into resonance. Therefore, they are called intervention 'addressees'.

Box 2: Definition of complexity in social systems theory

Complexity in social systems theory refers to contingency, operational closure and structure. First, the increase of the number of a system's elements results in the decrease of the probability that all relations between the elements can be realized. "Everything that goes beyond this is based on selection, thus generating contingent states (states otherwise possible)." (Luhmann 1997/2012: 79) Second, the elements are 'communications' and form different types of social systems: on the micro level interaction systems (face-to-face communication), on the macro level the overall (world) society with societal sub-systems (media guided communication), and on the micro-meso level organisational systems (communication based on decision programs). Third, operational closure of social systems allows for structural complexity ordering communications along 'symbolically generalized communication media SGC-media' such as 'money' in the case of the economic system, 'power' in the case of the political system, and scientifically proven 'truth' in the case of the science system (Luhmann 1988: 52; 1981/1990: 31, 43–48; 1990: 272). SGC-media together with

6 Any natural or legal person or public entity or group of such persons and/or entities, including any temporary association of undertakings, which offers the execution of works and/or a work, the supply of products or the provision of services on the market (OJEU 2014/L/94–24).

structures in the form of expectational nexuses such as 'persons', 'roles', 'programs', and 'values' provide various compounds of selection schemata (Luhmann 1984/1995: 315–319; 1997/2012: 50) – e.g. dominated by economic rationality (with the binary code payment/non-payment) or by political rationality (with the binary code government/opposition) or by one of the other societal rationalities (science, mass media, education etc.)[7].

In the case of the PPPI initiative, various actors can be identified performing a certain intervention role at the organisational level (Figure 1). Economic operators supplying innovative solutions upon public entities demands which, in turn, must receive intervention signals sent from innovation policy makers. These signals have to be tailored in a way which increases the probability that they are received by the public entities as information (and not as noise). In the terminology of social systems theory, resonance is achieved whenever signals in the environment affect the system's self-referential operation mode.

Since the political systems are differentiated into politics, administration and public (Luhmann 1981/1990: 47) three respective roles have to be considered. Simplified it can be said that politics means those entities which are disposed to change with elections (politicians), administration means those entities not disposed to change with election, but directly delivering public services (civil servants as policy makers with their agencies, guided by ministers & government and parliament), and the public means citizens as users of public services and as electorate. To make the PPPI initiative successful, political decisions and administrative execution must strengthen each other and eventually, public acceptance must be achieved. Political decisions describe and/or define the intervention goal as the difference between before and after the intervention; administrative execution has to design and implement intervention instruments to provide the adequate signals; and the public as end-users will afterwards be satisfied or not.

From all of the above mentioned actors, PPPI addressees and professional experts (analysts, evaluators, associations, mass media, etc.) will be the most attentive observers of the intervention subject's and intervention administrator's activities. But the latter and the public will also occasionally or continuously take care. All learnings should be documented and consolidated to provide the basis for the next round of the intervention cycle.

7 See, for the discussion of system rationalities and their innovation specific selectivity, Buchinger 2009a: 189–192.

Figure 1: Schematic PPPI Intervention Cycle

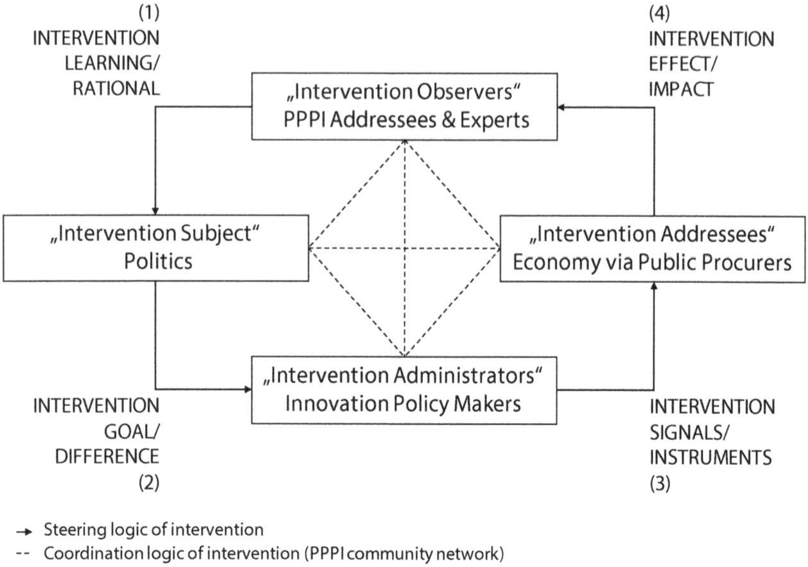

Source: adopted from Buchinger 2009c, 2007

4 Phases of implementation

4.1 First lab phase: Learning and clarifying the context

Austria's innovation procurement initiative has been triggered by the European Commission, which has itself been inspired by the US procurement model SBIR[8]. Innovation procurement became an explicit issue in Austria in 2007. An innovation procurement guide "procure_inno" (BMWA 2007) has been formulated and broadly disseminated. Although the guide was well received within the public procurement community, its mobilization effect remained modest because of missing accompanying measures. Therefore, several studies were executed on behalf of the Federal Ministry of Science, Research and Economy (BMWFW[9]) and the Federal Ministry for Transport, Innovation and Technology (BMVIT) to

8 SBIR stands for Small Business Innovation Research and is a set-aside program for small business to engage in federal research & development with potential for commercialization. See beside the above mentioned expert group reports, EC 2007/C/860, 2007, 2007/C/799; US-Congress 1982, 1992; Wessner 2007.

9 At that time (2007), Federal Ministry of Economy and Labor (BMWA).

learn from international good practices in policy making and to identify national/international innovation procurement good practice examples[10]. Beyond that, a dialog with representatives of the Austrian Action Plan Sustainable Procurement was initialized. All these activities resulted in the up-taking of the demand-side issue in the Austrian Strategy for Research, Technology and Innovation (RTI-Strategy) (BKA et al. 2011).

On the basis of the RTI-Strategy, the above mentioned ministries started to prepare the PPPI Lab. The first milestone was the mandating of a PPPI Action Plan by the Austrian government (BMWFJ and BMVIT 2011). As it turned out subsequently, the government mandate was quite important in motivating stakeholders to participate in the PPPI lab process.

4.2 Second lab phase: Co-designing and deciding policy intervention

In April 2011, a stakeholder process was launched to co-design the intervention as a crucial part of the development of the Austrian PPPI Action Plan. Process owners were (and still are) cooperatively the Ministry of Science, Research and Economy (BMWFW[11]) and the Ministry for Transport, Innovation and Technology (BMVIT). Before the stakeholder process was started, a stakeholder mapping took place. That is, a functional analysis was conducted to distinguish the various roles performed, identify thus the 'right' stakeholders and structure the process accordingly:

- Role 'intervention subject': politicians setting PPPI goals and mandating responsibilities
- Role 'intervention administrator': policy makers implementing PPPI instruments
- Role 'buyer as intervention addressee 1st tier': public entities demanding innovative solutions in their call for tenders
- Role 'user as intervention addressee 1st tier': buying public entity as users of the procured works/supplies/services
- Role 'supplier as intervention addressee 2nd tier': firms and other economic operators delivering innovative solutions as responding to public entities' demand

10 E.g. Buchinger 2009b; Buchinger and Pfliegl 2011; Buchinger and Steindl 2009; Clement and Walter 2010.

11 At that time (2011), Federal Ministry of Economy, Family and Youth (BMWFJ).

- Role 'end-user as intervention addressee 3rd tier': citizens benefitting from low-cost & high-quality services, from an empowering social environment and from a sound nature

The structure of the stakeholder co-design process is represented in Figure 2. Policy makers, public procurers, industrial suppliers and citizens' representatives were invited. The response to the invitation was very positive and more than 90 stakeholders participated; only citizens' representatives did not follow the invitation. The stakeholder process focused on the three issues 'policy instruments', 'procurement processes', and 'innovation conditions' installing a working group for each of them. They were accompanied by a further working group considering benchmarking/data issues. On the basis of the outcome of these four working groups, the framework for an integrated policy mix was elaborated in a discussion forum which resulted in a catalogue of desired PPPI interventions (support, activities).

Figure 2: Architecture, dates/mandates of the PPPI stakeholder process

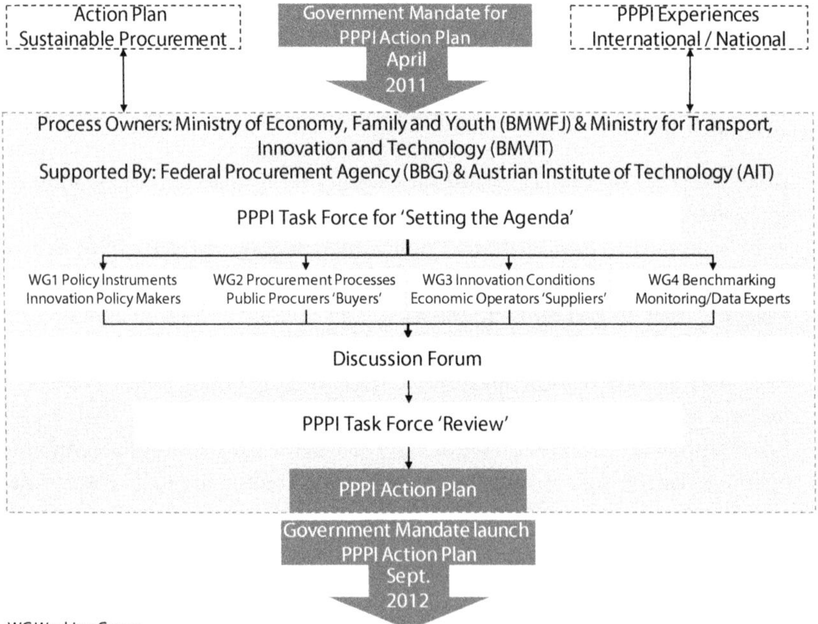

WG Working Group

Source: Buchinger 2012

The strategic framing of the stakeholder process was provided by the PPPI Task Force ('setting the agenda' & 'review'). Following the traditional Austrian social partnership principle, it included not only key procurers (e.g. state/provincial governmental bodies, Austrian Motorway and Expressway Corporation ASFINAG, Austrian Federal Railways ÖBB, Federal Real Estate Company BIG, Energy AG VERBUND, Federal Postal Services POST) and intermediaries (e.g. Austrian Research Promotion Agency FFG, Austrian Economic Service AWS, Austrian Standards Institute ASI), but also key federations (e.g. Austrian Economic Chambers WKÖ, Federation of Austrian Industries IV, Austrian Chambers of Labor AK, Vienna Hospital Association KAV) and leading firms. The operative responsibility was assigned to a steering group consisting of the process owning ministries (BMWFW, BMVIT) and experts (Federal Procurement Agency BBG, Austrian Institute of Technology AIT).

The Austrian PPPI Action Plan was successfully completed after 16 months and – as a second milestone of the PPPI lab – its implementation was mandated in September 2012 by the Austrian Government (BMWFJ and BMVIT 2012b). A set of activities was suggested which should provide adequate signals for the intervention addressees:

- *Strategic dimension* ("soft law"): integration of innovation needs of public procurers in existing programs; political commitment to the introduction of innovation-related procurement plans in public entities and dedication of budgets
- *Operational dimension* (state aid & procurement): establishment of PPPI service centres to provide tailored support; establishment of financial incentives stimulating innovation procurement; initializing and supporting innovation procurement projects
- *Legislative dimension* ("hard law"): amending procurement law to include innovation as an explicit issue
- *Impact dimension*: establishment of an innovation procurement monitoring and benchmark system

Since the measurement of innovation procurement was and is still difficult, the PPPI Action Plan does not specify a quantified target. Instead, the aim was generally formulated as "increasing the share of innovation related public procurement in Austria" (BMWFJ and BMVIT 2012a: 13).

4.3 Third lab phase: Intervention implementation and interim assessments

The implementation of the PPPI Action Plan started immediately after the government mandate was given. Up to date the following interventions took place and indicating all together the third milestone of the PPPI lab.

Within the *strategic dimension*, the integration of innovation needs of public procurers in existing programs has been successful in the program Mobility of the Future – Transport Infrastructure Research VIF (started 2011/12), is on its way in the program ICT of the Future (started 2016), and the opportunities to include them in the program City of the Future will be checked in 2017. There has been a dedication of budget for awareness, support and training activities for the period 2012–2017 and recently, a further PPPI budget dedication has been decided by the government for the period 2018–2021. The political commitment to the introduction of innovation-related procurement plans in public entities is an open issue till now.

Within the *operational dimension*, a PPPI Service Centre was initially established within the Federal Procurement Agency (BBG) in 2013. Subsequently, PPPI competence centres have been set up in order to cooperate closely with the service centre. At the moment these are: the Austrian Research Promotion Agency FFG (PPPI Service Partner focusing on pre-commercial procurement of R&D); the Austrian Federal Promotional Bank AWS (PPPI Service Partner focusing on commercial procurement of innovative solutions); the Austrian Federal Real Estate Company BIG (PPPI Service Partner focusing on the buildings sector); the Austrian Association for Transport and Infrastructure GSV (PPPI Service Partner focusing on the transport sector); the Austrian Energy Agency (PPPI Service Partner focusing on the energy sector); the Austrian Economic Chamber WKO and the Federation of Austrian Industries IV (PPPI Service Partners focusing on the exchange with the industry); and the Procurement Expert Conference of the Provinces (PPPI Service Partner focusing on the exchange with the provincial governments). They all work together in providing a service network and have been establishing the following services and financial incentives:

- Trainings for public procurers, networking opportunities (face-to-face and electronic), an information roadshow taking place in every one of Austria's provinces, and a biannual awareness event combining innovation and green procurement. These generic activities are complemented by the following specific provisions.

- Online market place which gives public procurers the opportunity to specify a need or problem as a challenge and suppliers the chance to present innovative solutions for commercial procurement.
- Awarding of vouchers on the basis of a contest which can be used by public procurers for PPPI-support such as technology consulting, legal advice, or project management support. The vouchers and the additionally offered consultation services for the preparation of PPPI projects are for free, being selected on the basis of a competition.
- Awarding of grants for pre-commercial procurement of R&D (PCP) in the thematic programs mentioned above.
- Awarding of grants for R&D with the mandatory inclusion of public procurers in the field of 'security' (KIRAS, started already before the PPPI Action Plan was developed and is now continued as part of the PPPI initiative).

The "PPPI Annual Report 2013/2014" includes detailed descriptions of the PPPI stakeholders, their activities and a series of results in the form of PPPI good practices (BMVIT and BMWFW 2015). The next PPPI report (2015/2016) is presently in its preparation.

Within the *legislative dimension*, the amendment of the Austrian procurement law of 2013 successfully includes innovation as an explicit issue (BGBl 2013/128). The draft of the reform of procurement law 2017 (which is the translation of the 2014 EU Procurement Directive into national law) has been sent out for review recently (BKA 2017).

Several interim assessments have taken place to intensify the learning processes within the PPPI lab process. Thereby, PPPI events and the work of the PPPI service network were investigated. The results, which were condensed in the mid-term review of the Austrian Strategy for Research and Innovation (BMWFW and BMVIT 2016), indicate that the PPPI initiative is well received by the Austrian procurement community up to now. A comparative analysis of Austria's innovation procurement performance in the EU and the OECD context (Buchinger 2016; ERAC 2015; OECD 2016) highlights Austria's strategic strengths on the one hand, and honours its rapid catching up vis-à-vis leading European countries such as the Netherlands and UK on the other hand. The Austrian Council for Research and Technology Development recommends specific activities to foster this transformation (Rat-FTE 2016).

4.4 Fourth lab phase: Monitoring & impact evaluation

The above mentioned interim assessments are part of the development of a comprising Austrian monitoring system for innovation procurement. Another part of the evolving monitoring system is the nationwide measurement of public innovation procurement, regardless of being stimulated by the PPPI initiative or not. This is a difficult task because of the lack of sufficiently defined indicators. Statistics Austria conducted therefore in 2014/2015 a pilot survey asking for the financial volume of innovation procurement in the year 2013. The result was of a range between 2.3 % and 3.3 % of innovation procurement as part of the overall procurement budget in the state sector[12]. There are no representative figures yet for the outsourced (public) enterprises. Since it is part of the ERAC recommendations to the European Council to "invite the Commission to develop the monitoring system of innovation procurement in the EU, including a new indicator 'innovation procurement' within the Innovation Union Scoreboard" (ERAC 2015: 3), Austria will coordinate its next steps with EU activities.

Concerning the PPPI initiative, a comprising impact evaluation is scheduled for 2017/2018. It will show how much and what kind of resonance has been achieved by the above described interventions. On this basis, it will provide recommendations for the further improvement of the PPPI initiative.

5 Conclusion and outlook

The PPPI lab – i.e. policy lab to design and conduct an intervention aiming at higher rates of public procurement promoting innovation – is meanwhile well received in Austria. Between 2011 and 2016 a step-by-step institutionalization process took place. Valuing the progress of institutionalization on the basis of the milestones (M) described in this text, Austria has been quite successful up to now. Following the logic of the intervention cycle, evidence for the political success is provided by M1 'mandating the development of a PPPI Action Plan by the Austrian government in 2011', and M2 'conducting the stakeholder co-design of the PPPI Action Plan 2011/2012 and receiving the government mandate for

---

12 Three types of innovation procurement have been included: 1) goods/services developed for the procurer (R&D), 2) goods/services were the procurer was the first buyer, 3) goods/services new on the market and new for the procurer. Uncertainties concerning PPPI categories became apparent in the pilot survey. Scenarios were, therefore, calculated with results for an innovation procurement share between 2.3 % and 3.3 % of investigated public entities procurement budget in the reporting year 2013. Source: Statistics Austria.

its implementation in 2012'. Evidence for the administrative success is provided by M3 'implementing measures such as the amendment of the procurement law, the inclusion of procurer needs in existing programs, the provision of a range of services and financial incentives resulting in concrete innovation procurements, and the initialization of a monitoring system between 2012 and 2016'.

Up to now it has been a laborious learning process for policy makers and their supporting agencies (PPPI service network) on the one hand and for public procurers on the other. However, the raised awareness, the growing use of the offered services and the resulting innovation procurements are rewarding the efforts.

Nevertheless, important work remains to be done. Above all, for the coming time it is necessary to move from a good practice 'elite' towards an innovation procurement 'mass-movement'. Or in the terms of the intervention logic, to move from a state where only a few early adopters are open to the signals of the PPPI initiative to a state where the broader innovation procurement community is receptive to these signals. The forthcoming impact evaluation (2017/2018) will guide further improvement of the PPPI initiative.

References

BGBl (2006/17). Bundesvergabegesetz 2006 (geltende Fassung 2017) [*Austrian procurement law*]. Bundesgesetzblatt für die Republik Österreich, Wien.

BGBl (2013/128). Novelle Bundesvergabegesetz (BVergG) und Bundesvergabegesetz Verteidigung und Sicherheit (BVergGVS) [*Amendment of Austrian procurement law*]. Bundesgesetzblatt für die Republik Österreich, Wien.

BKA (2017). Entwurf Vergaberechtsreformgesetz 2017 [*Draft Reform of procurement law 2017*]. Bundeskanzleramt, Wien.

BKA, BMF, BMUKK, BMVIT, BMWFJ and BMWF (2011). Strategy for research, technology and innovation of the Austrian Federal Government. Austrian Federal Chancellery, Vienna.

BMVIT and BMWFW (2015). IÖB-Jahresbericht 2013/2014 [*PPPI Annual report 2013/2014*]. Bundesministerium für Verkehr, Innovation und Technologie & Bundesministerium für Wissenschaft, Forschung und Wirtschaft, Wien.

BMWA (2007). procure_inno: Praxisorientierter Leitfaden für ein innovationsförderndes öffentliches Beschaffungs- und Vergabewesen [*Guide on public procurement promoting innovation*]. Bundesministerium für Wirtschaft und Arbeit, Wien.

BMWFJ and BMVIT (2011). Ministerratsvortrag: Entwicklung eines österreichischen Leitkonzeptes für eine innovationsfördernde öffentliche Beschaffung (IöB) [*Government mandate to formulate an action plan on public procurement promoting innovation*]. Bundesministerium für Wirtschaft, Familie und Jugend & Bundesministerium für Verkehr, Innovation und Technologie, Wien.

BMWFJ and BMVIT (2012a). Leitkonzept für eine innovationsfördernde öffentliche Beschaffung (IÖB) in Österreich [*Austrian Action Plan: Public procurement promoting innovation PPPI*]. Bundesministerium für Wirtschaft, Familie und Jugend & Bundesministerium für Verkehr, Innovation und Technologie, Wien.

BMWFJ and BMVIT (2012b). Ministerratsvortrag: Umsetzung des österreichischen Leitkonzeptes für eine innovationsfördernde öffentliche Beschaffung (IÖB) [*Government mandate to implement the action plan on public procurement promoting innovation]*. Bundesministerium für Wirtschaft, Familie und Jugend & Bundesministerium für Verkehr, Innovation und Technologie, Wien.

BMWFW and BMVIT (2016). Austrian research and technology report 2016. Federal Ministry of Science, Research and Economy (BMWFW) & Federal Ministry for Transport, Innovation and Technology (BMVIT), Vienna.

Buchinger, E. (2007). Applying Luhmann to conceptualize public governance of autopoietic organisations. *Cybernetics & Human Knowing*, 14(2–3), 173–187.

Buchinger, E. (2009a). Innovation as a complex societal process: System rationalities and professional roles. In: Aguado, J. M., Scott, B. and Buchinger, E. (eds.), Technology and social complexity. Editum University of Murcia, Murcia.

Buchinger, E. (2009b). Innovationsfördernde öffentliche Beschaffung: Innovationspolitische Optionen [*Public procurement promoting innovation: Options for innovation policy*]. Austrian Institute of Technology, Wien.

Buchinger, E. (2009c). Organisationen, Autopoiesis und Governance-Modi: Komplexitätsmanagement am Beispiel der Wiener Clusterpolitik [*Organisations, autopoiesis and governance-modi: Complexity management by the example of the Viennese cluster policy*]. In: Weyer, J. and Schulz-Schaeffer, I. (eds.), Management komplexer Systeme: Konzepte für die Bewältigung von Intransparenz, Unsicherheit und Chaos. Oldenburg, München, 215–234

Buchinger, E. (2012). PPPI Policy Notes: The Austrian innovation procurement strategy process 2011/12. Austrian Institute of Technology, Vienna.

Buchinger, E. (2016). PPPI Policy Notes: Austria's innovation procurement performance in the EU and ORCD context. Austrian Institute of Technology, Vienna.

Buchinger, E. and Pfliegl, R. (2011). Innovationsfördernde öffentliche Beschaffung: Handlungsoptionen im vorkommerziellen Bereich im Themenfeld Verkehr [*Public procurement promoting innovation: Options for pre-commercial procurement in the transport sector*] Austrian Institute of Technology, Wien.

Buchinger, E. and Steindl, C. (2009). Innovationsfördernde öffentliche Beschaffung: Ein neues Instrument der Innovationspolitik? [*Public procurement pro-*

moting innovation: A new instrument for innovation policy?]. Austrian Institute of Technology, Wien.

Clement, W. and Walter, E. (2010). Innovationsfördernde öffentliche Beschaffung in Österreich [*Public procurement promoting innovation in Austria*]. Foresee, Wien.

EC-Expert-Group (2003). Raising EU R&D intensity: Improving the effectiveness of the mix of public support mechanisms for private sector research and development. [Expert Group: Georghiou L./chair, Rigby J./rapporteur, Amanatidou E., Belitz H., Cruz L., Edler J., Edquist C., Granstrand O., Guinet J., Leprince E., Orsenigo L., Romanainen J., Stampfer M., Biesen J.v.d.]. European Commission, Brussels.

EC-Expert-Group (2005). Public procurement for research and innovation [Expert Group: Wilkinson R./chair, Georghiou L. and Cave J./rapporteurs, Bosch-Cantallops C., Caloghirou Y., Corvers S., Dalpé R., Edler J., Hornbanger K., Mabile M., Montejo M.-J., Nilsson H., O'Leary R., Piga G., Tronslin P., Ward E.]. European Commission, Brussels.

EC-Expert-Group (2006). Creating an innovative Europe: Report of the independent expert group on R&D following the Hampton Court Summit [Expert Group: Aho E./chair, Georghiou L./rapporteur, Comu J., Subirá A.]. European Commission, Luxembourg.

EC (2007). Guide on dealing with innovative solutions in public procurement: 10 elements of good practice. European Commission, Brussels.

EC (2007/C/799). Pre-commercial procurement: Driving innovation to ensure sustainable high quality public services in Europe. European Commission, Brussels.

EC (2007/C/860). A lead market initiative for Europe. European Commission, Brussels.

EC (2008). European code of best practices facilitating access by SMEs to public procurement contracts. European Commission, Brussels.

EC (2010/C/546). Europe 2020 flagship initiative innovation union. European Commission, Brussels.

EC (2016a). Buying green: A handbook on green public procurement. European Commission, Brussels.

EC (2016b). Single market scoreboard: Public procurement. European Commission, Brussels.

Edler, J. and Georghiou, L. (2007). Public procurement and innovation: Resurrecting the demand side. *Research Policy*, 36, 949–963.

Edquist, C., Vonortas, N. S., Zabala-Iturriagagoitia, J. M. and Edler, J. (eds.) (2015). Public procurement for innovation. Edward Elgar Publishing, Cheltenham.

EPA (2014). Green procurement: Guidance for the public sector. Environmental Protection Agency, Wexford.

ERAC (2015). ERAC opinion on innovation procurement. European Research Area and Innovation Committee (Task Fore Members: Laatsit M./rapporteur, Adolphe D., Beley B., Buchinger E., Ortiz Pajares M., Garrido M., Kalytis R., Pavlaković I., Rajalo S., Svendsen L., Vermeulen H., Vilén K., Voort L., Putten M.), Brussels.

Luhmann, N. (1981/1990). Political theory in the welfare state (Revised translation from the German original '*Politische Theorie im Wohlfahrtsstaat*' of 1981 by John Bednarz Jr.). De Gruyter, Berlin-New York.

Luhmann, N. (1984/1995). Social systems (Translated from the German original *Soziale Systeme: Grundriss einer allgemeinen Theorie* by John Bednarz Jr. with Dirk Baecker). Stanford University Press, Stanford.

Luhmann, N. (1988). Die Wirtschaft der Gesellschaft [*Economy as societal system*]. Suhrkamp, Frankfurt.

Luhmann, N. (1990). Die Wissenschaft der Gesellschaft [*Science as societal system*]. Suhrkamp, Frankfurt.

Luhmann, N. (1997). Limits of steering. *Theory, Culture & Society*, 14(1), 41–57.

Luhmann, N. (1997/2012). Theory of society Vol. 1 (Translated from the German original 'Gesellschaft der Gesellschaft' of 1997 by Rhodes Barrett in 2012). Stanford University Press, Stanford.

OECD (2010). The OECD innovation strategy: Getting a head start on tomorrow. Organisation for Economic Co-operation and Development, Paris.

OECD (2011). Demand-side innovation policies. Organisation for Economic Co-operation and Development, Paris.

OECD (2014). Intelligent demand: Policy rationale, design and potential benefits. Organisation for Economic Co-operation and Development, Paris.

OECD (2015). Going green: Best practices for sustainable procurement. Organisation for Economic Co-operation and Development, Paris.

OECD (2016). Public procurement for innovation: Good practices and strategies. Organisation for Economic Co-operation and Development, Paris.

OJEU (2014/L/94–24). Directive 2014/24/EU: Public procurement. Official Journal of the European Union, Brussels.

OJEU (2014/L/94–25). Directive 2014/25/EU: Procurement by entities operating in the water, energy, transport and postal services. Official Journal of the European Union, Brussels.

Rat-FTE (2016). Empfehlung zur innovationsfördernden öffentlichen Beschaffung in Österreich [*Recommendation on public procurement promoting innovation in Austria*]. Rat für Forschung und Technologieentwicklung, Wien.

SBIR-STTR-IPC (2014). Report to the congress: SBIR/STTR outreach. Small Business Innovation Research & Small Business Technology Transfer Program lnteragency Policy Committee.

SocialPlatform (2015). Public procurement for social progress: A Social Platform guide to the EU Public Procurement Directive SocialPlatform, Brussels.

US-Congress (1982). Small business innovation development act. Senate and House of Representatives of the United States of America, Washington.

US-Congress (1992). Small business research and development enhancement act of 1992. Senate and House of Representatives of the United States of America, Washington.

Wessner, C. W. (ed.) (2007). SBIR and the Phase III Challenge of Commercialization. The National Academies Press, Washington.

Barbara Streicher, Margit Fischer, Wolfgang Czerny,
Barbara Heller-Schuh

The Austrian ScienceCenter-Network as an example for systemic intervention in science learning

Abstract: *The Austrian Science Center Network serves as an experimental setting for putting network theory into practice. It aims at stimulating systemic interventions in various societal subsystems by fostering hands-on science learning. The paper analyses the structure, instruments and outcomes of this organisation during its first decade.*

1 Introduction

If scientists, museum educators, artists, teachers, science communicators and policy makers engage in lively discussion and open exchange even at the 58th network meeting – then something must have worked out in the Austrian ScienceCenter-Network. If 170 partners from such diverse institutional backgrounds focus on their common active interest in hands-on science learning and, without formal commitment, stay engaged and willing to learn and cooperate – then the network must provide some value to them. If the initiators of the Network[1] look back at the first 11 years – they smile. In retrospect, it seems straightforward to put network theory into practice for a systemic intervention – however, it started as and still is an ongoing experimental process, requiring careful observation and governance. This paper intends to analyse the structure, instruments and outcomes of this unusual organisational form.

For many of the challenges our society is facing, the attitudes we have towards learning as well as towards science and technology are key elements for change. The underlying idea of science centers and their activities is that self-determined, playful and interactive engagement with science and technology can spark fascination and interest, can be empowering to their users and can foster a joy in learning. If these prospects hold true, we have to make use of the full potential of

1 i.e. the founding board members of the non-profit Association ScienceCenter-Netzwerk: Margit Fischer, Josef Fröhlich, Wolfgang Czerny, Monica Stadler, Barbara Streicher.

science center activities for societal benefit. This touches on the needs of several societal subsystems: The innovation system needs (young) people interested in science and technology careers from all socioeconomic, educational or ethnic backgrounds as well as gender. In the education system, the shift from knowledge to competences goes along with a need for self-determined and practical learning experiences. In all areas of our society, being able to critically question and understand, to engage in dialogue and to reflect on science and technology issues that affect our lives, are important competences and elements of a mature democracy.

Table 1: Key questions driving the Association ScienceCenter-Network

How can we inspire young PEOPLE to pursue science or technology careers?
How can we support joyful LEARNING for all ages?
How can we tap into the POTENTIAL of all people in a globalised science and economy?
How can we foster SOCIAL INCLUSION in a fragmented society?
How can we induce DIALOGUE about complex topics?
How can we value SCIENCE as part of our culture?

Table 2: Science Center Activities – as defined within the Austrian network

engage laypeople so they may experience and understand scientific topics, technological phenomena and connections between them
– are interactive (hands-on or minds-on)
– enable self-determined learning
– do not presume prior knowledge
– have a playful component
– give impulses to think further

We believe that open learning environments, as provided by science center activities, can positively contribute to these challenges. This led us to look for a matching organisational form that in itself would represent an open learning environment, which supports insights about and broader implementation of science center activities. Stimulating such implementation processes in the Austrian learning landscape requires an understanding of complex systems.

2 Structure

2.1 Principles of complex systems

In a complex system, the components interact in a way that is non-trivial and thus cannot be predicted by standard linear equations. Self-organisational processes of its actors, as well as small variations in initial conditions or (external) interventions, will influence the time-dependent changes in the state of the system. Its overall behaviour can thus only be understood as an emergent consequence of all embedded behaviours, as the whole is more than the sum of its parts (Levy 1994).

Social Systems Theory (based on Niklas Luhmann 1984) identified three key principles that are at work in complex systems: self-organisation, self-production and requisite variety.

The capacity for self-organisation is a property of complex systems which enables them to develop or change internal structures spontaneously and adaptively in order to cope with or manipulate their environment (Cilliers 1998). This principle implies an openness of the system, so that it exchanges resources (energy or material) with its environment.

Self-production (or autopoiesis) is the capability of a system to reproduce and maintain itself (Maturana 1980). It develops own behaviour, rules and communication patterns, which shape its identity. In complex networks, governance is a matter of autonomous self-control and not top-down steering from a central position (Little 2001). This principle implies autonomy of the system to keep it stable and alive.

Requisite variety is necessary for a system to deal productively with perturbations, e.g. from the environment. The larger the variety of actions that is available to the system, the larger the variety of perturbations it is able to compensate (Ashby 1958). This principle implies that diversity of actors in a system is needed to be optimally prepared for any contingency.

Networks are ideal models for the description of complex systems, as they show similar structures and features. An autopoietic social network is as a set of organisations or individual actors which interact non-trivially and come from various societal subsystems. Although those subsystems remain autonomous, the social network has the potential to orient and transform them through the self-organisational and reorientation processes of its actors that are stimulated through participation in the network.

In terms of organisational learning that fosters social innovation, networks have been found appropriate and useful structures (Willke 2004). However, they need a social entrepreneur taking the risk of founding and operating reflection spaces in order to pool actors within "cultural islands" (Schein 2010) focusing on social innovation (Wilhelmer 2012). Such guided networks are able to become "change agents" within civil societies by offering free space for entrepreneurship, organisational and network learning. They build on system dynamics and illustrate that complex organisations have no single optimum, but different optima in different times (Forrester 1971).

3 Setup of the Austrian ScienceCenter-Network

For the idea of stimulating interactive hands-on learning in the Austrian (science) learning landscape – which touches on subsystems such as education, science, innovation, museums, arts, policy, economy – initiating a social network seemed a valid experimental approach: it could generate the intended impact and at the same time serve as a practical test of network theory.

The Association ScienceCenter-Network Austria was set up mid-2005 as a non-profit entity. In January 2006, the proponents invited relevant actors to a first meeting, sounding their needs and ideas as well as developing a common vision for the to-be-established network. Immediately, a number of organisations committed themselves and became the first partners of the network.

Inside this (since largely grown) network, the Association fulfils a number of functions which are critical with respect to the steering of the network, also keeping the above-mentioned principles of openness, autonomy and diversity in focus.

Firstly, it operates as a hub that provides orientation and organisational functions and serves as a contact point within the network, for partners as well as for individuals and institutions from outside. Over the years, the Association has also gained international reputation and is a well-respected entry point into the Austrian science engagement and education community.

A second function is creating learning environments, thus deliberately providing interventions that can mobilise and orient the self-organisation processes of partners. Such interventions include activities, events, projects and processes that invite participation, as well as documentation, research and trainings that invite reflective practice.[2] The Association reacts to weak signals that show potential and then initiates interventions or irritations, which can be regarded as experimental triggers that might or might not be taken up by the partners. It looks out

2 The Association also undertakes projects and research on its own. The guiding principles for these are not to act in competition with partners, but to go for edgy activities that push the field and to care for the overall benefit and learning of the network by providing access to the results and processes that have been developed.

for a resonance to issues and interventions within the network as well as beyond, including financial donors, in order to let go of or to reinforce topics and issues.

The third function of the Association is network governance, thus continuously and reflectively observing the development of the system, which in this context is regarded as the network, its actors as well as its environment, both nationally and internationally. This includes listening into discussions, observing trends, conducting social network analysis, as well as developing and testing instruments and formats within the network.

3.1 Network governance

Orienting and supporting a social network relies on a set of characteristics that can shape network culture. Among these are ensuring cooperation, confidence, commitment, reliability, negotiation, sustainable partnerships and systematic insights. Within the Austrian ScienceCenter-Network, the Association actively uses this advice deduced from network theory to build the principles and instruments within the network.

From the start, the Austrian ScienceCenter-Network proved to be a structure unfamiliar to many actors in the field. It is not a membership organisation, not a federation nor an umbrella organisation. Instead, it is a deliberately non-hierarchical structure, even though the Association assumes a central role. All partners maintain their independence and self-organisation and they do not have to pay any fees. Partnership, however, does require a commitment of the respective partner, for genuine interest in science center activities, for openness in sharing, as well as for participation in activities of the network if adequate. An unusual, but proven feature of the network is that partners are not actively recruited for the sake of completeness of actors. Hence, only truly interested individuals or organisations seek out partnerships and thus account for an active network.

This special structure in itself influences the culture within the network. It remains an open structure, as partners themselves define their interest and participation as adequate. This is well reflected in the resulting high diversity of partners, with regard to their expertise[3], size, as well as their geographical location. The structure enables fairness as every partner is considered of equal value, no matter if it is an individual person, a small organisation or a large institution, differences are being respected. Giving and taking are important values, as the benefit of being a network partner is dependent on his/her engagement and sharing. No formal

3 The more than 170 partners can be grouped into categories of museums, (science) communication, research, education, economy and arts & media.

decision-making processes are required within the network, yet, the Association listens closely to the upcoming issues and trends, with a view to the overall benefit of the network. This allows great flexibility and the ability to react quickly to new ideas, opportunities and requests. At the same time, it requires trust, commitment and participation from everyone who wants to shape the course of the network.

The structure of the network thus corresponds closely to its main content focus: Just as science center activities provide a rich environment for self-determined personal learning, facilitated by explainers, so the network, facilitated by the Association, provides a rich learning environment for self-organisation of partners. In both, active engagement based on individual preferences and conditions is the key to learning, making connections and change processes.

A special governance element, unique to this network, is the very active involvement of a chairperson with high societal reputation, Margit Fischer. Being the wife of the Austrian President[4], her prestige greatly supports the confidence and reliability ascribed to the network, while her long-term engagement and expertise in the science center field, coupled with her open and modest personality, exemplifies commitment and communication style for the network.

4 Instruments

4.1 Network meetings

Frequent network meetings are the most important instrument within the Austrian ScienceCenter-Network. Those regular physical encounters – which are possible due to the limited size of the country – form the basis for connections between actors, for a common understanding of issues and goals around interactive hands-on learning as well as for establishing network culture.

During the first 11 years, 57 network meetings took place, each with an attendance of 40–80 participants. All network meetings are open to guests (usually 10–20 % of participants), ensuring the openness that is required for self-organisation of a social network.

The character of ScienceCenter-Network meetings also reflects the principles of open and interactive learning that are at the heart of science center activities, which are the common interest of partners. Characteristic features of the network meetings thus include an appealing setting, interactive, sometimes playful formats, small group discussions and professional moderation. Hence, engaging with the topic and with each other feels essential and natural for participants,

4 Dr. Heinz Fischer, Federal President of the Republic of Austria 2004–2016.

they easily develop connections and confidence. The now well-established format that supports open-minded exchange and learning also serves as an element of reliability and commitment within the network. By carefully documenting the issues and discussions of network meetings, the Association ensures that even those partners that could not participate feel involved and informed about network developments.

Network meetings take place five times a year, twice a year outside of Vienna, usually at a partner's venue. This is an opportunity to visit each other and foster peer-to-peer feedback on activities or exhibitions, strengthening confidence and the sense of a learning community. Partners who offer to host a network meeting influence the chosen topic, they show their commitment and thus may become attractive partners for cooperation.

Determining a topic for a network meeting is one of the options for intervention within the network. By highlighting particular issues, the Association provides an impulse and at the same time monitors the resonance of partners with the particular issue. Most topics arise through discussions within the network, by the suggestion of a partner or by scanning national and international trends in the fields of informal learning and science communication. A number of meetings have focused on network development, ensuring systematic insights and reliability. The initial meetings also included important elements of negotiation, resulting in the definition of "science center activities" (see Table 2) and a mission statement for the network (Science Center Netzwerk 2007).

The network meetings build the ground for inventions in the social network that can be seen as emergent phenomena, as predicted by network theory. Among these are project cooperations, events, trainings, invitations, publications, etc., where partners are interacting bi- or multilaterally with each other, often to the benefit of a specific target group or audience.

4.2 Cooperative projects

Initiating cooperative projects inside the network is another important instrument that allows orientation of actors which might then influence their respective (sub)system. By openly inviting participation, partners and other interested organisations can be mobilised to develop content on a specific topic, engage with a particular target group or become more attentive to a certain attitude or principle.

While ideas for projects are often invented and creatively played with inside the network, efficient implementation requires a different structure. Thus, organisationally, a project has to be separate from the regular network activities, with a hierarchical structure and clearly defined goals and roles. Ensuring a balance and clear distinction between open network processes and efficient project structures proved to be one of the most important learning steps during the development of the network.

Projects and cooperations are emergent phenomena of the Austrian ScienceCenter-Network. They utilise the complementarity of partners' expertises and the trust built up within the network. Large innovative and cooperative projects often involve the Association in the function of project management; additionally, numerous other cooperations have been formed between various subsets of network partners.

Travelling exhibitions are examples of network projects that demonstrate well the scope and vitality of the Austrian-wide network. The three exhibitions developed so far were conceived in network meetings, negotiating suitable topics and formats. Eventually, partners were invited to contribute with their interactive exhibits, which were curated and organised by the Association and a project team into a coherent, modular and robust travelling exhibition. All three were characterised as highly interdisciplinary, hands-on and engaging exhibitions[5]. Partners could participate by providing exhibits, working as a member of the project team (design, text, facilitation, etc.) or by hosting the exhibition in their venues. They profit from having their work shown across Austria, thus gaining visibility, from access to a highly engaging exhibition as well as from reflective discussions around the project and its results. A number of participating organisations were inspired by the project to develop and build their first interactive hands-on exhibits and learned from support and feedback by their peers. Obviously, the largest impact is on the more than 150.000 visitors of the exhibitions, among them a large number of school groups. As accompanying research showed (e.g. Hossein 2009), the interactive, interdisciplinary, open setting was highly attractive and inspiring for students and teachers.

Another type of project intervention includes concerted activities throughout Austria. These mobilise partners to offer public activities on a certain topic during a specific awareness week, sometimes coupled with international events[6]. Publishing this as a network activity strengthens the cohesion and identification with the network and provides an incentive for taking up new topics. An impressive example of a coordinated activity of the Austrian ScienceCenter-Network was the

5 Erlebnis Netz[werk]e / Discovering Net[work]s (2007–2010); Grenzgenial / Borders and boundaries (2009–2013); Wirkungswechsel / Interdependencies (2014-ongoing); for details on all projects, see www.science-center-net.at

6 Topics for action weeks were, e.g., polar melting, space, mobility or planet under pressure.

participation in the 1st International Science Center and Science Museum Day in November 2016[7]. For this event, hosted in the Natural History Museum Vienna, hands-on activities on each of the 17 UN Sustainable Development Goals engaged visitors in discovering the science and scope of action around these global topics. Again, this event mobilised partners as well as the exchange with the surrounding subsystems, including sustainability actors that became aware of the interactive approaches the network has to offer. Visitors from schools and the general public were highly satisfied and enjoyed the open format that invited self-guided and playful exploration and triggered interesting discussions on the complex global topics.

The opportunity to host the Annual Conference of the European Science Center Network ECSITE in 2016 in Austria[8] was a strong moment of identification for the network. The three hosts[9] invited the other networks partners to shape the conference, especially with a Nocturne event showcasing the Austrian ScienceCenter-Network to the international community. Additionally, a large public event[10] invited network partners, schools, research teams, companies and passers-by to participate in constructing elements for a huge cooperative chain reaction machine. Not only were partners proud to be part of this, they also learned a new format of engaging the public. As the Association ScieneCenter-Network had already gained experience with chain reactions in various other settings[11], it since provides support for partners wanting to transfer the activity into their own specific contexts (e.g. in museums, schools and public events), thus widening access of the public to such engaging activities, inspiring learning for all ages.

Further important project formats included workshops in schools, discussion games as well as action research projects. For example, understanding of inquiry-based teaching in primary schools was fostered when six network partners worked for a year with schools in their vicinity, probing how interactive science learning was taken up by pupils and teachers. The results regarding activities and research were disseminated in the network as well as to the educational system with pub-

---

7 In 17 Zielen um die Welt / Around the world in 17 goals (November 10th, 2016).

8 ECSITE is the European network for science centers and science museums. The conference took place in Graz, June 7–11 2016.

9 Childrens museum Graz FRida & freD, Universalmuseum Joanneum, Association ScienceCenter-Netzwerk.

10 „Ecsite for all“: more than 90 palettes with diverse chain reaction mechanisms were built, resulting in a 13 minute chain reaction.

11 In holiday workshops with kids & parents, during a walk-in action week, inside the knowledge°room, etc.

lications, an extensive report and a symposium (Frantz-Pittner 2011, IMST 2010, Science Center Netzwerk 2010a, 2010b).

A special project is the “Knowledge°room“[12], which transforms empty shops in underserved areas of Vienna into pop-up science centers, reaching out to non-privileged communities, including migrants and refugees. Partners can use these as experimental spaces to test activities and formats for audiences that rarely visit their museums or activities. This project not only stimulated reflection about social inclusion (see below), it also motivated transfer of the project idea into further contexts. Since 2013, the knowledge°rooms successfully attracted thousands of visitors, many of them from low socioeconomic and migrant backgrounds, serving as a place where empowerment can happen through engagement with science.

4.3 Working groups and transfer

With the steady growth of the network, a need for differentiation became obvious. Thus, a new instrument for more intense discussions evolved in the form of working groups on specific subtopics. Coordinated by the Association, partners meet regularly in order to advance and spread the issues, thereby providing products and processes for the relevant societal subsystems.

For example, the network has worked on developing a training scheme for science communicators (explainers), which has already been put into practice in various seminars and attracted participants from research, museums, universities, education and science communication[13].

With a view to the formal education system, partners have commonly devised a model development plan (Bertsch 2014) for Austrian schools who want to put a focus on science and technology, supporting their self-guided change processes. Schools were also provided with guidelines and resources on how to foster STEM learning (science/technology/engineering/mathematics) and how to make use of the informal science learning field.

A series of meetings on social inclusion in science communication managed to further attract participants from arts communication as well as relevant gatekeepers and members of migrant communities who are interested in building contacts with cultural and science institutions. Apart from great discussions, valuable contacts and individual cooperations, a result for the public was the development of

12 Wissens°raum / knowledge°room, since 2013 in Vienna, awarded with Sozialmarie Prize for social inclusion 2014.

13 Impulse and advanced seminars on explainer professionalisation, since 2013.

a website for socially inclusive science and arts communication[14]. Currently, this group has joined forces with the European science center community to work on a framework for social inclusion in science centers and museums.

These examples clearly show how the network, with its partners from various societal subsystems, is able to identify, process and condense issues of importance. The results not only influence the partners' orientation, but through raising interest and deliberate transfer also reach further actors and subsystems. This is especially valuable as changes in systems are only possible by bundles of synchronised measures. Reorientation of the actors inside these systems, each with limited influence, can add up to larger societal effects.

4.4 Public relations and documentation

Visibility of the network and its work is essential for effecting changes in the learning landscape. As instruments for public relations, an extensive web platform, as well as regular newsletters to different audiences, serve to provide information on partners, on science engagement and learning opportunities. More and more, social media like Facebook have become an important medium for exchange and references to current developments and offers, connecting the community also virtually.

Publishing brochures as newspaper attachments is a means of communicating highlights of network and partner activities to a broader public, emphasising yearly topics that could be interesting for societal uptake.

Documentation of network meetings, project and study results, as well as formats and instructions for well-tried hands-on science center activities, is another valuable form of sharing knowledge as well as broadly spreading the principles of informal science learning.

4.5 Research, teaching and consultancy

The network is a learning community. Thus, encouraging reflective practice and supporting professional development within the network and beyond is a constant concern of the Association. An important instrument is conducting or contracting research on specific interactive formats and activities, as well as on their impact on target groups. A number of studies and accompanying research projects have been performed (e.g. Schütz 2009, Hossein 2010, Unterleitner 2010, Schulze 2012, Gruber 2016) and results are being shared in network meetings, workshops or through reports.

14 www.gemmahin.at, since Jan 2017

Teaching and trainings, offered to partners, schools, museums and innovators, have become a significant means of dissemination about informal learning and hands-on science center activities. This has a strong multiplier effect with a potential influence on the formal education system as well as on science communication as intended by research institutions and companies. Most efficiently, such seminars are offered in cooperation with established teaching and training organisations like (pedagogic) universities or human resource departments.

Recognising their expertise and overview with regards to the Austrian science learning landscape, representatives of the Association have been invited as speakers, discussants and consultants on these issues both nationally and internationally. Again, this expertise is rooted in the constant exchange with partners and careful observation of the relevant subsystems and provides an opportunity for impulses that might stimulate re-orientation processes.

5 Outcomes

5.1 Network development

In the 11 years since its foundation, the network has grown to more than 170 partners Austria-wide. While in 2005 no dedicated science center existed in Austria, in 2016 a number of organisations were explicitly using the term "science center" to describe their offers[15]. Additionally, numerous other partners work with science center activities, yet chose a different name. Among the partners are not only institutions, but also individuals with a high interest in the network that focuses on science center activities, some wishing to remain partners even after leaving their institutions.

Diversity is high inside the network, as it comprises:

- partners from the science center, museum and science communication field, for which science center activities are a central element of their work;
- partners from research organisations and the economy, wishing to communicate their results and processes in a more understandable, hands-on, engaging way;
- partners from educational institutions like schools, kindergartens or teacher training organisations, who appreciate the offers of informal learning locations and the ideas, methods and resources provided by science center didactics;

---

15 Inatura (Dornbirn), Haus der Natur (Salzburg), EXPI Science Center (Gotschuchen), Experimentarium (Hartberg), Welios Science Center (Wels), Audioversum (Innsbruck).

- partners from arts, design and culture, who work on the border of science and arts;
- partners from the media, who communicate about science (learning).

The overall development of the network has been described as five phases: preparation, build-up, development, reflection and diversification phase (Wilhelmer 2011, 2012). In the first five years, two external counselors[16] with expertise in systems theory and organisational development supported the conceptual design, meta-reflection and co-facilitation of network meetings with dialogue-oriented methods, thus supporting a "soft government" approach and training on the job for the network executive manager (Wilhelmer 2012).

Financing of the fundamental network functions as performed by the Association ScienceCenter-Network as well as for specific projects has been provided by public sources (federal ministries, provinces, cities), social partners and industry associations, by companies, donors, as well as through European projects.

5.2 Network analysis

In order to gain systematic insights as a basis for network governance, social network analysis has been periodically conducted to assess existing connections between partners. The quantitative analysis focused on questions about the structure and density of the network as well as the correlations between the positioning of partners and their activity level within the network. A first insight was gained in 2006 (Heller-Schuh 2006) followed by a comprehensive network analysis performed in 2008/09 (Wenk 2009) and again with a new data set and analysis in 2015/16 (Heller-Schuh 2016).

In accordance with the principle of matching content and format, even the collection of data took the form of a science center activity with reflective potential. Each partner received a set of (80 in 2008, 161 in 2015) cards with all other partners, being asked to sort them into known/unknown and send their established contacts back with comments on the intensity of their interaction. This unconventional and engaging method not only sparked discussion in the partners' teams, it also provided high return rates of 81 % and 74 % (in 2008 and 2015, respectively).

In total, nearly 3.000 connections between individual partners were identified, whereby more than 1.300 were classified as cooperations, characterised within this study as having a benefit to third parties.

16 Doris Wilhelmer, Petra Wagner (Austrian Institute for Technology).

The results of a typical network visualisation are shown in Figure 1. Each node represents a partner, whereby size corresponds to the number of contacts and colours to the organisational type. The position of a node within the visualisation is dependent on its connection to all other nodes, with highly connected actors tending to be more central.

Figure 1: Social network analysis – Visualisation of the ScienceCenter-Network in 2015

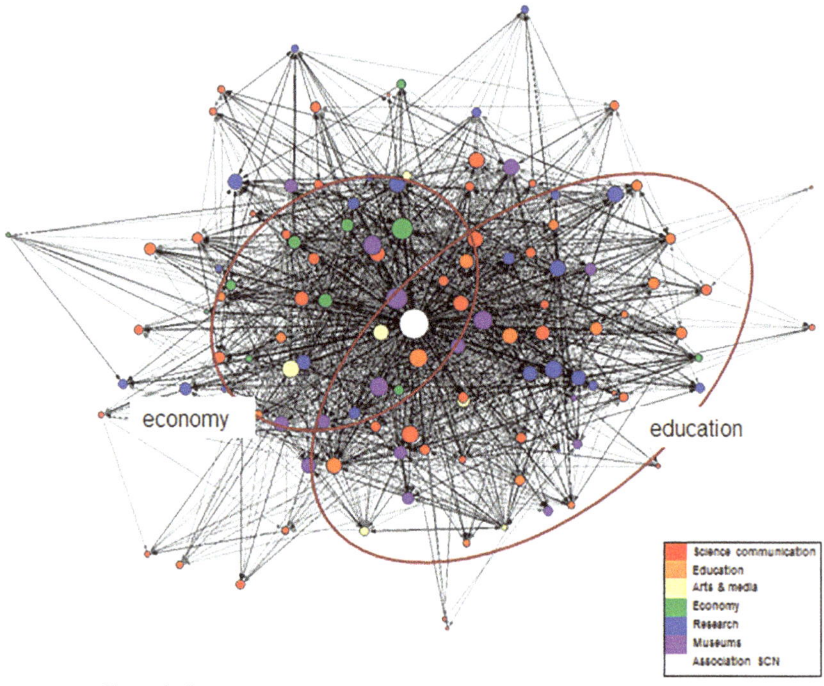

Source: Heller-Schuh 2016

In 2015/16, the Austrian ScienceCenter-Network was characterised by high connectivity and density, with short distances between partners (2–4 nodes) and an average of 20 connections per partner. Unsurprisingly, the Association assumes a central role; however, the network would not be fragmented without this node. There are more connections between actors of the same geographical region, consistent with regional clustering. Partners of similar organisational background also tend to have stronger connections – especially among partners in education and the economy, but less for research, science communication and museums, who seem to serve a bridging function.

Strong correlations were identified between positioning of a partner and his/her activity level in the network. Actors that regularly attend network meetings, participate in projects and trainings or host events show the strongest connections. Multivariate statistical methods and regression analysis confirmed a significantly higher probability for contacts, cooperations and projects among actors if they attended the same network meeting (Heller-Schuh 2016), corroborating the central function of this important instrument for emergent network phenomena.

The analysis of the 1.300 co-operations that were specified further showed a broad range of activities that partners are engaging in with each other: inviting each other as experts for lectures and workshops, interviews and common text production; commissioning a contract; conducting research together; producing an exhibition; cooperating in an activity or event; co-participating in a network project or travelling exhibition; and even forming long-term structural co-operations between partners (Heller-Schuh 2016).

Overall, social network analysis of the ScienceCenter-Network proved the positive effect of several instruments introduced through network governance, both for establishing the weak couplings of knowing one another and for activating them into strong connections of active cooperation.

In 2008/9, a qualitative study was undertaken in parallel to the quantitative network analysis, whereby the researcher interviewed all (then 80) partners of the network. Important findings were the high satisfaction of partners with the network and its offers and benefits for them. Some reported an initial unfamiliarity (and sometimes uneasy feelings) with the non-hierarchical structure of the ScienceCenter-Network. By gaining confidence and showing commitment, however, partners came to regard the flexible structure and open culture as a particular asset of the network. The high diversity of partners was often stated as the most exciting element of the ScienceCenter-Network, allowing fruitful discussions and an insight into different perceptions on issues of common interest. The work of the Association ScienceCenter-Network was commented as central to the drive and development of the network, and as an essential, professional and trustworthy actor with a high commitment to the well-being of the whole system (Wenk 2009).

An important learning outcome for the board and team of the Association was the need to be more precise about the distinction between the network and the Association, as the two share the same name and content focus, yet have very different organisational and decision structures.

The results of the network analyses were shared with the network partners, enabling them to assess their own position by means of ego-networks that highlight their own connections[17], thus further stimulating self-reflection, as well as to gain insight into systems theory and learning.

5.3 Orienting the network – an example

Among the many issues where interventions were deliberately introduced to trigger self-organisational and re-orientation processes, the topic of social inclusion can serve as a model to describe the effects of the various instruments described above.

Listening to partners and the international discussion was the starting point for the issue of social inclusion to be raised within the network, when a partner shared inspiration from an international conference within a network meeting. Immediately, the Association recognised both the potential of this weak signal and the interest the topic received among the participants. Thus, it dedicated a network meeting (and two more since) to the issue in order to create awareness and sensitivity for the issue.

As an opportunity for experiences, the project "knowledge°room" was initiated and partners were invited to contribute to the project with content and to use it as an experimental space. From the beginning, this project was accompanied by a social scientist who performed research on the social dimension and the effects of the initiative (Gruber 2016). An event with an open space workshop followed by a panel discussion was set up to publicly discuss the potential that the low-threshold, hands-on science communication offered in local "knowledge°rooms" may have for social inclusion, community work, education and economy. As facilitators and role models in the pop-up science centers, young people with mixed ethnic background are employed, actively using the diversity of competences present in our communities.

A second line of involvement for partners in socially inclusive activities is the long-term cooperation with the Vienna youth prison, which regularly invites network partners to conduct hands-on workshops for their inmates. Additionally, a series of excursions has been offered by network partners to students from the University of Economy who serve as learning buddies for socially deprived youth. Each of those opportunities is disseminated by the Association to all partners, thus periodically reinforcing the idea of social inclusion as a possible focus for their work.

To foster professional development, international experts were invited to present their strategies to empower disadvantaged target groups from migrant com-

---

17 Available online at http://www.science-center-net.at/netzwerk_verbindungen/.

munities or low socioeconomic background through engagement with science center activities. A two-day workshop, again with international colleagues, offered professional development for facilitators and partners wanting to extend their skills regarding social inclusion.

The project "knowledge°room" is even internationally one of the few successful attempts of socially inclusive science communication. Thus, results and insights were shared at conferences and events[18] and received high international interest and acknowledgement in the form of invitations for publication (Streicher 2014a, 2015a) as well as participation[19]. In 2014, the project was entered in a competition and was awarded with the "Sozialmarie", a prize for social innovation. This recognition, as well as asking for funding for the project and thereby illustrating its impact, can be seen as a form of input into the Austrian education and innovation system.

Realising that migrant groups could be more effectively addressed through respected mediators from their own communities, a series of special meetings was initiated. Here, science communication professionals, museum staff, community representatives and non-profit organisations supporting migrants and refugees meet to discuss experiences and ideas on how science and culture programs could be structured and advertised in order to be attractive and socially inclusive. As a result of these meetings, a number of museums and science organisations have initiated cooperations with a cultural translator and/or with community groups, offering special and low-threshold activities for non-privileged groups. A new web platform[20] conceived by this working group intends to bundle and publicise these offers.

The idea of socially inclusive science engagement is also taken up – or the timing coincides with societal developments – by public bodies: Funding calls for science communication which specifically ask for a socially inclusive dimension of submitted projects[21]. As one of the successful submissions, the "knowledge°room" idea is currently being transferred into a different context[22], with counselling by

18 e.g. Ecsite conferences (2014, 2015), Noyce Leadership Program (2013–2014), EUCU-NET conference (2014), Forum Wissenschaftskommunikation (2015).

19 e.g. in a science festival in Mersin, Turkey (2015).

20 www.gemmahin.at.

21 Support program of the Austrian Ministry for Transport, Innovation and Technology „Talente regional. Kinder, Unternehmen und die Welt der Forschung". Calls 2015/16 put a focus on migrant and diverse social backgrounds.

22 Project "Schau rein!", ARGE WIKI, URL: https://www.ffg.at/sites/default/files/images/programmlinien/kurzbeschreibung_schau_rein.pdf.

the Association, who is eager to further learn from the experiences with adaptation and to feedback the insights to the whole network and its environment. The Vienna business agency explicitly asked the ScienceCenter-Network for trained facilitators to support the city science festival 2015 in their efforts to address and facilitate visits for refugees.

5.4 Reference point and model structure

Within the Austrian education and innovation system, the Association ScienceCenter-Network as the hub of the network has become a reference point and an important input provider. Not only is the term "science centers" now being used as a specific term within public documents, also the network is specifically referred to as an expert organisation and learning community. The team of the Association is frequently invited by media, projects and public bodies as experts for working groups, panel discussions and written contributions on issues around science communication, science policy and education (e.g. Streicher 2014b, 2015b, 2016).

Additionally, the Austria ScienceCenter-Network has reached a prominent place in the worldwide science center landscape and is regarded as a reliable international partner for cooperations. Its special organisational form is seen as a prototype and interesting model structure for an organisation with very low fixed costs, but a high outreach, diversity and flexibility that is envied by conventional science centers. The network experiment is being carefully studied by international partners (e.g. Becker 2015) and, upon consultancy by the Austrian Association about the structure, experiences and underlying network theory, also being transferred into other contexts[23].

6 Outlook

After more than a decade, the Austrian ScienceCenter-Network is a success story, yet it also remains an ongoing experimental process. With regards to the initial idea of the network serving as a practical test of network theory and social system theory (based on Luhmann 1984), the observations confirm that the key principles of openness, diversity and autonomy are actively cared for by the Association and valued by partners. The advice from network theory with regard to important elements of network governance is being considered and the effect of some of

23 e.g. Young explorers' clubs, which are organised as a network in Poland – http://www.kopernik.org.pl/en/workshops/young-explorers-club/.

the instruments that have been established based upon it was confirmed by the results of network analysis. Emergent projects and monitoring of developments show an impact, or at least resonance with partners, on visitors and relevant societal subsystems.

The future of a complex system is not predictable. Shaping the future requires a constant and vigilant re-evaluation of internal ideas, goals and processes as well as monitoring of how the overall system develops and a willingness to experiment, learn and adapt.

We will end with two sets of questions that seem relevant to the future of the network:

How can the ScienceCenter-Network keep the balance between being an established and well-trusted organisation and still keep the spirit and momentum of a constantly experimental mind-set? Will it be possible to take the learning network to a next level, by offering it a physical space for reflective practice, without losing the key principles of openness, autonomy, diversity and flexibility?

With the rise of virtual networks with their random nature and constant state of flux, how will the ScienceCenter-Network keep the commitment of partners focused on science center activities? Which instruments will prove essential for stressing the open, yet intentional nature of the network that serves as the basis for cohesion, cooperation and combined strength towards the common goal of stimulating interactive hands-on science learning?

Acknowledgements

The idea and structure of the Austrian ScienceCenter-Network would never have been possible without the expertise and considerateness of Prof. Josef Fröhlich, co-founder of the Association. We would also like to thank Petra Wagner and Doris Wilhelmer for their support in shaping and analysing the development of our organisation.

References

Ashby, W.R. (1958). Requisite variety and its implications for the control of complex systems. Cybernetica, Vol 1, issue 2, pp. 83–99 (available at http://pcp.vub.ac.be/Books/AshbyReqVar.pdf, republished on the web by F. Heylighen – Principia Cybernetica Project).

Becker, J. (2014). Mapping european science engagement networks. Spokes magazine. ECSITE. URL: http://www.ecsite.eu/sites/default/files/spokes5_mapping_networks.pdf [accessed Feb 2, 2017].

Bertsch, C. (2014). Forschend Lernen im naturwissenschaftlich-technischen Unterricht in der Grundschule. URL: http://www.science-center-net.at/fileadmin/SCN_new/Projekte/SQA/SQA_Forschend_Lernen_Grundschule_FINAL.pdf [accessed Feb 2, 2017].

Cilliers, P. (1998). Boundaries, Hierarchies and Networks in Complex Systems. International Journal of Innovation Management, Vol 5, issue 2, pp. 135–147.

Forrester, J. (1971). Counterintuitive behavior of social systems. Technology Review 73(3): 52–68.

Frantz-Pittner, A., S. Grabner and G. Bachmann (eds.) (2011). Science Center Didaktik. Forschendes Lernen in der Elementarpädagogik. Hohengehren: Schneider Verlag.

Gruber, S. (2016). Zusammenfassende Analyse zum Wissens°Raum am Dornerplatz (1170 Wien). ScienceCenter-Network, unpublished.

Heller-Schuh, B. (2006). Berechnung und Visualisierung des Science Center Netzwerks. ScienceCenter-Network, unpublished.

Heller-Schuh, B., M. Dünser and M.-E. Züger (2016). Visualisierung und Analyse des ScienceCenter-Netzwerks III. URL: http://www.science-center-net.at/index.php?id=738 [accessed Feb 2, 2017].

Hossein, S., B. Streicher, M. Bachler (2009). Begleitforschung zur Ausstellung „Erlebnis Netz(werk)e" am Standort Wien http://www.science-center-net.at/fileadmin/SCN_new/Downloads/Presse/download/ENW_Endbericht_Begleitforschung.pdf [accessed Feb 2, 2017].

Hossein, S., V. Wieland, M. Kreuzer, K. Unterleitner and B. Streicher (2010). Begleitforschung zum Projekt KET – Kinder erleben Technik. URL: http://www.science-center-net.at/index.php?id=476 [accessed Feb 2, 2017].

IMST, Otto Schütz, ScienceCenter-Netzwerk (eds.) (2010). „Außerschulische Lernorte". IMST Newsletter Jahrgang 9, Ausgabe 33. Institut für Unterrichts- und Schulentwicklung (IUS), Alpen-Adria-Universität Klagenfurt.

Levy, D.L. (1994). Applications and Limitations of Complexity Theory in Organisation Theory and Strategy. Available from: https://www.researchgate.net/publication/268060010_Applications_and_Limitations_of_Complexity_Theory_in_Organisation_Theory_and_Strategy [accessed Feb 4, 2017].

Little J. H. (2001). Autopoiesis and governance: Societal steering and control in democratic societies. In: Geyer F. & van der Zouwen J. (eds.) Sociocybernetics. Complexity, autopoiesis, and observation of social systems. Greenwood Press, Westport CT: 159–170.

Luhmann, N. (1984). Soziale Systeme: Grundriß einer allgemeinen Theorie. Frankfurt: Suhrkamp.

Maturana, H., F. Varela (1980). Autopoiesis and Cognition: the Realization of the Living (1st edition 1973, 2nd 1980). Dordecht: Reidel Publishing Co.

Science Center Netzwerk (2007). Leitbild. URL: http://www.science-center-net.at/index.php?id=215#c1198 [accessed Feb 2, 2017].

ScienceCenter-Netzwerk et al. (2010a). Forschend lernen: Partnerschaften zwischen Volksschulen und Science Center Einrichtungen. URL: http://www.science-center-net.at/fileadmin/SCN_new/Projekte/ForschendLernen/Forschend_Lernen_EB_1_Allgemeiner_Teil_Final_20100505.pdf [accessed Feb 2, 2017].

ScienceCenter-Netzwerk (2010b). Dokumentation des Symposiums Forschend Lernen und Partnerschaften – ein Innovationsmotor. 9. – 10. November 2010 im Kassensaal der Oesterreichischen Nationalbank in Wien, Dezember 2010. URL: http://www.science-center-net.at/fileadmin/SCN_new/Projekte/Symposium/A-DOKUMENTATION_Symposium_FL_FINAL.pdf [accessed Feb 2, 2017].

Schein, E. H. (2010). Organisational culture and leadership (Vol. 2): John Wiley & Sons.

Schulze, H., C. Längle and A. Natmessnig (2012). Mobil mit Hirn. Forschungstudie zur Umsetzung von Schulprojekten zum Thema „Intelligente Mobilität". Eine Studie im Auftrag des Bundesministeriums für Verkehr, Innovation und Technologie. URL: http://www.science-center-net.at/fileadmin/SCN_new/Projekte/Mobil_mit_Hirn/Endbericht_Mobil_mit_Hirn.pdf [accessed Feb 2, 2017].

Schütz, O., U. Kozeluh and J. Müller (2009). Good Practice Elemente von dialogisch/ diskursiven Verfahren und niederschwelligen Science-Center-Aktivitäten zur Unterstützung von Good Governance im Bereich Wissenschaft und Gesellschaft. Studie im Auftrag des Rat für Forschung und Technologieentwicklung. URL: http://www.science-center-net.at/index.php?id=415 [accessed Feb 2, 2017].

Streicher, B., K. Unterleitner and H. Schulze (2014a). Knowledge rooms – science communication in local, welcoming spaces to foster social inclusion. JCOM 13 (02)(2014)C03.

Streicher, B. (2014b). Neugier braucht kein Vorwissen. Gastkommentar APA Science Dossier. URL: http://science.apa.at/dossier/Neugier_braucht_kein_Vorwissen/SCI_20140227_SCI39692494217208188 [accessed Feb 2, 2017].

Streicher, B. (2015a). Do Science Centers Need Buildings? Informal Learning Review ILR July/August 2015, 29–30.

Streicher, B. (2015b). „Wissenschaft – Gesellschaft: eine (Neu)Orientierung." In: Wissenschaft und Gesellschaft im Dialog. „Responsible Science", Bundesministerium für Wissenschaft, Forschung und Wirtschaft (Hg.), 53–57.

Streicher, B. (2016). Science Center und ihre Aktivitäten als Schnittstelle zwischen Universität und Erwachsenenbildung. In: Magazin erwachsenenbildung.at. Das Fachmedium für Forschung, Praxis und Diskurs. Ausgabe 27, Wien.

Unterleitner, K., S. Gruber and B. Streicher (2010). Grundlegende Charakteristika und Prinzipien für den Dialog zwischen Wissenschaft und Gesellschaft. Studie im Auftrag des Rat für Forschung und Technologieentwicklung. URL: http://www.science-center-net.at/index.php?id=486 [accessed Feb 2, 2017].

Wenk, B., B. Heller-Schuh, T. Scherngell, U. Oberhammer and B. Streicher (2009). Netzwerkanalyse 2008/9. URL: http://www.science-center-net.at/fileadmin/SCN_new/Forschung/Netzwerkanalyse_2008_9.pdf [accessed Feb 2, 2017].

Wilhelmer, D., H. Erler, P. Wagner and B. Streicher (2011). Uncertainty require management and counseling learning. Conference proceedings 5th MCD-Conference Amsterdam; 9th-11th June 2011; VU University Amsterdam.

Wilhelmer, D., H. Erler, P. Wagner, B. Streicher (2012). NETWORK Setups Diving Social Innovation. ZSI Discussion Paper, Nr. 29.

Willke, H. (2004). Einführung in das systemische Wisensmanagement, Heidelberg: Carl Auer Systeme Verlag.

Manfred Tscheligi

Technology Experience: A framework for people driven innovation

Abstract: *Technology Experience offers a conceptual and methodological framework to understand the perspectives, needs and wants of people towards existing and upcoming technology. The human experience and its essence have to be understood as a very important innovation driver. The article discusses three dimensions of Technology Experience relevant to technological changes that matter to individuals, organisations and society.*

1 Motivation

We don't just use technology; we live with it (McCarthy and Wright 2004). Much more deeply than ever before, dealing with technology involves us emotionally, intellectually, and sensually. It has become clear that there are a variety of qualities within a product, technology or service going beyond the pure functional aspects. Technological innovations are meant to work for people and their problems, rather than offer new challenges.

Science and innovation have started to talk about the notion of "experience" as a core ingredient of technological development. Apart from functionality, technology should create experiences that are meaningful and emotionally valuable for people. This is a switch from a pure instrument-oriented view of technological developments towards a broader viewpoint that includes aspects like pleasure, aesthetics, emotions, trust, acceptance, comfort, perceived safety or sociability. Experience-driven development has been evolved into a prominent strategy towards future innovative and successful technologies, which have to be embedded into different contextual situations and inherently include a high level of user experience consideration.

The experiential aspects of product use motivated several researchers to study user experience (UX) in different contexts and domain areas. Although UX is rooted in ICT, it is certainly not restricted to it. This is firstly caused by the fact that ICT pervades all dimensions of the "lifeworld" (living/working, personal mobility, health care, etc.). Secondly, this is due to the growing understanding that the experience dimension is critical to (more or less) all products/systems/services provided nowadays.

Experience universally covers everything personally encountered, undergone, or lived through. User experience is that subset of general experience which explicitly refers to the experience(s) derived from active/passive interaction with goods/services or a specific technology. As a result, several terms/concepts can be distinguished: utility, usability, user experience, customer experience and societal experience (see Figure 1).

Figure 4: From utility to experience

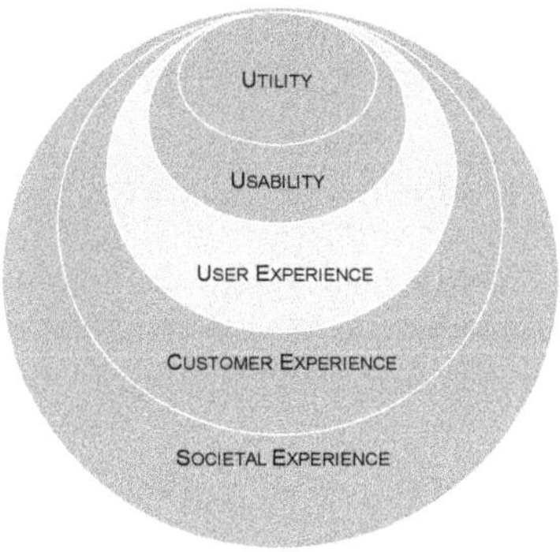

Utility: functionality of the system (product, service)
Usability: how well users can use system functionality, task and work-related (i.e. performance, smooth interaction, etc.), objective measures such as task completion time, number of clicks and errors, etc.
User experience: unique to an individual, influenced by and based on prior experiences and expectations, rooted in a social and cultural context, subjective measures such as a person's perception of good or bad, emotions, etc.
Customer experience: social and cultural context of UX in the form of lifestyle-criteria, such as brand, image, marketing, price, etc.
Societal experience: social and cultural context in the form of ethical issues, cultural traditions, grand challenges, etc.

Dealing with the optimization and exploitation of the diverse facets of experience at individual as well as organisational level is crucial for future technological innovation and adoption. As a consequence of digitisation, individualization and lifestyle orientation, experience is becoming a major paradigm in the industry, as well as in research and technology innovation. Considering innovation also, or firstly, from the experiential perspective will change priorities as well as maturity.

Specific application areas and application situations comprise specific interaction contexts, which leads to the more comprehensive concept of contextual experience. Recently technological innovations and new digital approaches have consistently been emerged in a variety of (complex) contexts. Along the technological diversification and contextualization (e.g. see the current discussions around Factories of the Future and Industry 4.0. as only one example), understanding this context from a people and usage centred perspective is crucial. The field of Technology Experience moved beyond the desktop and explored new forms of interaction in different contexts. Various theories and models motivating context-oriented thinking were proposed, such as the theory of situated action. We are now talking about Digital Contexts of the Future to capture the synergy of digital technologies and their specific use in a specific situation.

An in-depth understanding of contextual characteristics and the focused orientation of future experience approaches are a prerequisite for mature application technologies. This demands an extended viewpoint of user experience towards dedicated and specific contextual experiences. Even more essential is to understand the relationship between contextual specifics and experience needs as a basis for creating appropriate experience approaches driven by a well-defined experience quality framework. To name only one specific approach, ethnographic investigations offer a rich source for understanding and reflection.

Experience linked to specific *Contexts* are the two primary dimensions of this framework. Both come along with enhanced forms of Experience Oriented Thinking (EOT), EOT embodies a more radical way of experience oriented innovation. The wants and needs from an experience oriented perspective will be answered first, before technological developments will happen. EOT includes a very comprehensive and factorial way of dealing with the notion of experience. EOT will therefore act as the third dimension in our framework.

In the following sections these three dimensions will be discussed in more detail. We will introduce the basic concepts and what they deliver to technological changes that matter to individuals, organisations and society.

1.1 Characterising context

Over the past years the field of Technology Experience has moved beyond the desktop and has been exploring novel forms of interaction in different contexts. Various theories and models motivating context-oriented thinking have been proposed, such as the situated action theory suggesting that context determines how people behave in particular situations (Suchman 1987). An in-depth understanding of context enables application designers to choose what context to use in their applications (Dey 2001). A contextual interface will be designed according to the needs and behaviours of people in specific contextual situations and has the potential to be conceptually as well as technically adapted to the characteristics of the specific context.

The essence of situated action (Suchman 1987) is that experience is changed by the context in which it occurs. Moreover, combining specific contextual situations with deeper insights on user experience opens up new roads for research and challenges in all design and development phases. It will provide less obtrusive and more natural ways of interaction, resulting in advanced contextual interfaces.

A variety of laboratory and field study methods and techniques are already available, as well as different Context Labs have been developed over the last years in order to use them as experimental or real-live settings for studying, prototyping, and evaluating new concepts, designs or systems in an almost naturalistic environment. Some examples of context labs are the Aware Home Research Initiative (AHRI) at Georgia Institute of Technology, the Philips HomeLab, Care-Lab, Shop-Lab (Westerink et al. 2008) at Philips Research, the Intel PlaceLab, the Restaurant of the Future Context Lab developed by Wageningen University or the Fraunhofer inHaus project. In addition, so-called Living Labs are in development to study the incorporation of new technologies and applications in real life settings.

The term "context" has manifold meanings within various academic disciplines. Bradley and Dunlop (2005) provide a multidisciplinary overview on the understanding and modelling of context. In HCI, several definitions of context have been proposed during the last years (e.g. Shilit and Theimer 1994, Rodden et al. 1998, Lieberman and Selker 2000). One of the most comprehensive definitions of context is provided by Dey: "Context is any information that can be used to characterize the situation of an entity. An entity is a person, place, or object that is considered relevant to the interaction between a user and an application, including the user and applications themselves" (Dey 2001). From the interaction point of view, Dourish (2004) describes context as a relational property between objects or activities rather than as pure information. This means, context is actively produced, maintained and enacted in the course of the activity.

From a theoretical point of view, not only the situated action theory (Suchman 1987) has to be mentioned, it is also important to mention the Hutchins' theory of distributed cognition (Hutchins 1995), stating that processes take place distributed between groups of individuals and their environments (see also Massumi 2002). Moreover, the activity theory (see e.g. Nardi 1996, Engeström 1999) underpins the interdependence between people and technology. Kaenampornpan and O'Neill (2004) expanded the activity theory into a model of context by adding the time component. As with many other classifications of context, the level of granularity is either not clear or varies strongly with different components. Even within one particular model, different levels of granularity can be identified (Raptis et al. 2005). Moreover, the problem arises that some factors are easier to measure (e.g. position) than others (e.g. culture). The context classifications and models are also necessarily dependent on the specific application domain.

There are many different factors that constitute a particular situation, however the problem of identifying the relevant ones still remains when exploring a new context. Thus, many discussions take place around the factors comprising a specific context. Several researchers have tried to identify context factors, create context frameworks and models for different application areas (e.g. Dey and Abowd 1999, Dey 2001, Dey et al. 2001, Shilit and Theimer 1994, Petrelli et al. 2001, Pascoe 1998, Schmidt 2001, Hall 1976, Becker and Nicklas 2004, Jung et al. 2005, Raptis et al. 2005, Ho and Intille 2005, Hinze and Buchanan 2005, Venkatesh et al. 2003, Mihalic et al. 2006). A reflection can also be found in Murer et al. 2015.

In sum, the understanding of context towards the detailed relevance for designing interaction innovations and the notion of context is a slippery one and underpins the need for ongoing and empirically grounded research. More empirical evidence on the influencing factors and its relationships is needed. Deeply situated Contextual Labs are needed as key research and innovation mechanisms in synergy with industrial entities and practices to investigate the phenomena of contextual interaction (human-machine and human artifact interaction in a specific complex environment). The real environment is the lab and the lab plays back into the reality in a very reflective and research through design led form (Zimmerman et al. 2007).

1.2 User experience in context

The qualities of a product go beyond the instrumental aspects of product use and usability (Karapanos et al. 2008, Buxton 2007, Hassenzahl 2006). This has been recognised by user experience research, but the community is still looking for an adequate and overall definition of what constitutes "user experience". The concept of seeing technology in terms of experience was originally introduced by McCarthy and Wright (2004) and several attempts have already been made to describe the concept of user experience (Forlizzi and Battarbee 2004, Arhippainen and Tähti 2003, Hassenzahl 2003, 2004). Moreover, interaction experiences between people and artifacts can be investigated for different contexts on the visceral, behavioural and reflective levels (Norman 2004). For collecting data about user experience, it has become clear that a combination of quantitative and qualitative methods is needed (Tullis and Albert 2008).

The understanding of contextual experiences has significant potential for shaping a user's interaction with complex interactive systems. "In understanding how interaction shapes experience, time plays an important role in consciousness and in designing actions so that users like not just the outcome of the actions but the feeling of executing them“ (Ardito et al. 2007). Based on previous research on user experience, a selection of specific factors is considered as relevant for considerations in contextual situations.

Motivation depends on individual differences and on the environmental circumstances and how a person experiences the surrounding (Heckhausen 1980). As an example already in 1959 Herzberg (Herzberg 1959) proposed the Motivation-Hygiene Theory claiming for relevant motivator factors in factory working conditions like achievement, recognition, the work itself, responsibility, promotion and growth.

Emotions can be explained as a motivating force in action (Dewey 1980, Jordan 2000) and as a significant part of users' experiences. The main question is, how an interface can encourage positive emotions by also considering contextual information. We will also take the concept of (emotional) attachment, where we are working on, into account, as it refers to the strong connection a person feels towards a product/system for instance (see Bowlby 1969; Belk, Jacko and Sears 2003; Meschtscherjakov et al. 2014).

Fun/Enjoyment is an important factor, because in certain environments, people are very willing to be entertained. Brandtzæg and Følstad (2003) describe aspects of enjoyment, building on a demand-control-support model for good and healthy work. Csikszentmihalyi's (1990, 1997) flow approach is also useful to address fun/enjoyment aspects.

Sociability extracts the serious substance of life, leaving only “togetherness”, the sheer pleasure of the company of others (Simmel 1949; see also related co-experience approach of Battarbee 2004). Also (social) awareness, the “understanding of the activities of others, which provides a context for your own activity” (Dourish and Belotti 1992), is important to consider. “Awareness is essential in collaboration to coordinate the activities in a shared workspace“ (Wisneski et al. 1998) such as in production contexts.

User Acceptance can be defined as “the demonstrable willingness within a user group to employ technology for the tasks it is designed to support” (Dillion 2001). Davis et al. (1989) describe perceived usefulness (PU) and perceived ease of use (PEOU) as important factors. Venkatesh et al. (2003) further proposed the Unified Theory of Acceptance and Use of Technology (UTAUT).

Trust is recognised as an important factor since Kaasinen (2005) extended the TAM acceptance model by adding trust and ease of adoption, which are very important when introducing new interactive systems into a context, such as Human-Robot systems into the work context, into the factory environment. Trust includes perceived reliability of the technology and the provider, reliance on the service in planned usage situations.

To understand the ***Feeling of Security***, it is critical to understand how our brains work, and how they fail (Schneier 2008). By taking the feeling of security into account for the design of a system, in particular of new interfaces for the context car and factory, the user – driver/operator – should feel safe in using it.

Comfort relates to how comfortable users feel, in particular when new interaction approaches, such as ambient systems or Human-Robot Interaction are deployed. Exploring the concept "comfort" is not an easy activity, although the subjects are very familiar with the assessment of their subjective comfort level as investigated by Koay et al. (2006).

Sensoric Attitudes depend on everything humans perceive and that enters their minds through their senses. Designing for smell, taste, touch, sound and sight is combined in the term sensorical design (Shedroff 2001). A deeper understanding of sensorical attitudes can result in more intuitive and easy to grasp interfaces by interacting with users in an innovative way.

The above listed user experience factors show a limited collection of possible factors and need to be further extended and investigated in special contextual circumstances in order to understand the aspects that make an experience more successful and valuable to users and contribute to positive contextual user experiences. More factors will arise during contextual investigations and reflection.

1.3 Contextual experience methodologies

For studying prototyping and evaluating contextual experience, several methods and techniques have been developed and adopted. Several methodological approaches exist to study contextual influences on workflows and interaction paradigms and built a basis for design implications (Dourish 2006). Karen Holtzblatt et al. (2005) developed the methodological concept of Contextual Inquiry, which puts designers and engineers directly in the customers' work context, for gathering rich in-depth data about working routines.

As already mentioned ethnographic methods are field research methods that combine several data-gathering methods like participant observation, formal and informal interviewing and the analysis of documentary sources (Powdermaker 1966, Wax 1971, Werner and Schoepfle 1987). Ethnography provides detailed insight into people's behaviour, even if they themselves are unaware of this. Using ethnographies (Fetterman 1998) has become increasingly prominent within HCI (e.g. Blomberg et al. 1993, Crabtree 1998).

Cultural probing provides another way of gathering information on users, their activities and experiences. Compared to the previously mentioned field methods, it limits the effects of researchers on the users, as no researcher takes part in the daily life. Cultural probing allows users to self-report and is very useful to apply in an early phase of a user-centred process (Gaver 1999). Participants use a package of materials (e.g. post-cards, maps, photo albums or media diaries) to describe their everyday experiences and to answer questions in an informal way.

A different approach for measuring context is the Experience Sampling Method (ESM) (Larson and Csikszentmihalyi 1983). ESM is an in-situ method for collecting information about the context and the content of the daily life of individuals. It aims at catching emotions, motivations and cognitive processes when they actually occur in a real live situation (Mihalic et al. 2007, Meschtscherjakov et al. 2008).

There is also a growing interest in finding new ways to involve and engage users in the early design process of new applications and services, in particular addressing users in their real life context. Participatory methods and techniques attempt to involve users in the early idea generation process, right from the beginning of the requirement investigation and the following conceptual design process (Muller 2003). Co-design in particular means actively integrating end users in the development process of a future product by putting them in the role of designer and stakeholder (Obrist et al. 2008).

Prototypes help to illustrate and evaluate interaction concepts and interfaces. They differ in terms of functionality and performance, which depends on their intention of use (Arnowitz et al. 2006). Buchenau et al. (2000) introduce the term "experience prototyping", which "is a very dynamic, complex and subjective phenomenon. Buxton (2007) proposes to sketch early versions of a system or an artifact in various forms in order to inform the development team, to communicate ideas and to enable an early proof of design concepts. Research also showed the usefulness of the Wizard-of-Oz method (Nielsen 1993) in synergy with experience prototyping (Schrammel et al. 2007).

Visualizing and simulating the role of context, like the conceptual bond between context and subjects, persons, locations as well as composing context and role of artifacts are important to be visualized and simulated. In particular, context aware systems deal with modelling, specification, acquisition, interpretation and management of context (Chow 2006). Simulating the different ingredients of a contextual interface offers the possibility to investigate the user experience of a specific interaction innovation in the lab taking contextual conditions into account.

The role of evaluation is mainly to inform the design and to improve it at all stages of the design life cycle (Dix et al. 2004, Stone et al. 2005). Different kinds of evaluation may be carried out at different design stages and for different reasons. Evaluation done during the early design stages is undertaken to evaluate the users' requirements, predict the usability of the system, and to assess how well the interface meets the users' needs. Evaluation in a later stage is carried out to check the usability and user experience of a nearly complete system. Several methods and techniques like usability testing in laboratory environments or other in-situ methods, like experience sampling or cultural probing (as already described in the previous sections), can be used. Especially in the field of mobile experience methodological differences depending on the application of the method in a laboratory setting or in-situ are described in Bernhaupt et al. (2008). Baillie and Schatz (2005) in particular report the evaluation of multimodality in the laboratory and the field.

While the concept of patterns was developed by Christopher Alexander in urban architecture (Alexander et al. 1977), Bayle et al. (1998) claim that there is a need for new ways in order to deal with the increasing complexity and diversity of the interaction design field. Results revealed from contextual research within an initial analysis phase, during the design and development process and within evaluation and measurement are becoming more valuable and sustainable when they are presented in a more generalizable form. Contextual user experience patterns are an important topic. Patterns help to reduce complexity in a very well defined way.

1.4 Experience oriented thinking

In an increasingly competitive environment, innovation has become a driving factor for market leaders, technology leaders and hidden champions in a variety of industries. Not only has it become more important to innovate resourcefully, but also, to have better insights into existing and potentially upcoming needs and requirements of current and upcoming user groups at an earlier stage of product and service development. These demands for novel ways of measuring and anticipating the relevance of potential products and services are rising (Leavy 2010).

Experience Oriented Thinking (EOT) as Innovation Paradigm is fuelled by exactly that need for methods to understand structure and harness knowledge about user needs and aims to develop new innovation methods, which integrate a concrete outlook on the acceptance of future products and services in the markets into the innovation process. Linked to the understanding of contextual experience (experience in existing and upcoming future complex digital contexts)

EOT is equipped with methods to analyse and anticipate user needs, models the acceptance of future technologies and services and offers appropriate means to institutionalise these approaches.

Experience-oriented strategies within organisations demand appropriate processes, methods and tools. These have to be implemented within the organisations in the right way. Current internal mechanisms have to be redesigned under the regime of a user and experience-driven mindset. Organisations of all types have to be made aware of the potential of well-defined means instead of arbitrary activities. The Technology Experience Framework is ready to provide answers to these implicit or explicit demands.

In large and medium-sized companies Experience Oriented Thinking as Innovation Paradigm creates questions of strategic importance for company development and culture. "Companies need a "human-centred" design approach to navigate the blurring of lines between product and service, producer and consumer." (Brown 2009). This "human-centred" approach includes all aspects of experience and will have an impact not only on the company's portfolio of products and services, but also on development processes, management decisions, as well as employee behaviour and satisfaction.

In particular, start-ups in the era of digitisation will benefit by internalising Experience Oriented Thinking at a very early stage of their development of products/services and company culture. Here, a toolset of start-up-friendly technology acceptance models is a valuable methodological help for companies in early stages. A high level of contextual experience insights and experience maturity of their products and services are key for further steps and getting investors on board.

Other mechanisms relevant for EOT are contextualized acceptance models, which have the potential for prospective consideration of new and digital service and product ideas towards user viewpoints. These are based on approaches like TAM and extensions (Venkatesh & Davis 2000), the TAM3 (Venkatesh & Bala 2008), or the Unified Theories of Technology Acceptance Model (UTAUT) that unifies different models of technology acceptance (Venkatesh et al. 2003).

Putting the focus away from desktop-based information systems, the first work in adapting the technology acceptance model was carried out by (Lowry 2004), who focused on the cultural differences in technology acceptance. Relevant examples on the contextualisation of acceptance models were delivered to the automotive domain by (Meschtscherjakov et al. 2009) and the CTAM approach (Osswald et al. 2008). Contextualised acceptance models are a very important part of the Technology Experience framework as the offer anticipation first without building all aspects of the solution from an experience oriented viewpoint.

1.5 Implications and conclusions

Technology Experience offers a conceptual and methodological framework to understand the perspectives, needs and wants of people towards existing and upcoming technology. Technology experience delivers qualitative and quantitative insights and is able to capture the very essence of technology use. Technology experience offers powerful means towards anticipation and the experience driven construction of future technologies. The human experience and its understanding have to be understood as a very important innovation driver as most technologies are made directly or indirectly for the exploitation of humans in various application and demand contexts.

Experience is a highly context dependent phenomenon. Actual technology use and appropriation differ between contexts and are influenced by the various characteristics of a specific instance and different instances. Uncovering the relationships between experience factors and context factors is a research agenda of its own and by far not fully understood. Uncovering this complexity requires a wide set of technology studies in real and complex contexts combined with iterative design and reflection approaches. Bringing new technological artifacts into a given context changes the context itself. Research and innovation procedures very much demand this highly iterative nature of investigation. As an example, studying the requirements of people towards future technologies needs a rather multidimensional and design-led approach, rather than the practices of pure question and answer. People have to be confronted with futures as soon as possible to get feelings of needs and wants. This will be crucial as we are facing the challenges of future digital contexts. Technology Experience will be of help to answer fundamental questions of digitisation and what it will be.

References

Alexander, C., Ishikawa, S., Silverstein, M., Jacobson, M., Fiksdahl-King, I., Angel, S. (1977). A Pattern Language: Towns, Buildings, Construction, University Press, Oxford, UK.

Ardito, C., Costabile, M., Lanzilotti, R., Montinaro, F. (2007). Towards the evaluation of UX, In: Workshop "Towards a UX Manifesto" at the International Conference HCI 2007, Lancaster, UK, September 3, 2007, 6–9.

Arhippainen, L., Tähti, M. (2003). Empirical Evaluation of User Experience in Two Adaptive Mobile Application Prototypes, In: Proceedings of the International Conference on Mobile and Ubiquitous Multimedia, Norrköping.

Baillie, L., Schatz, R. (2005). Exploring Multimodality in the Laboratory and the Field, In: Proceedings ICMI'05, ACM Library, New York, USA, 100–107.

Battarbee, K. (2004). Co-Experience, Dissertation of the University of Art and Design, Helsinki, Finnland.

Bayle, E., Bellamy, R., Casaday, G., Erickson, T., Fincher, S., Grinter, B., Gross, B., Lehder, D., Marmolin, H., Moore, B., Potts, C., Skousen, G., Thomas, J. (1998). Putting it all together: towards a pattern language for interaction design. CHI 1997 Workshop.

Becker. C., Nicklas. D. (2004). Where do spatial context-models end and where do ontologies start? A proposal of a combined approach, In: First International Workshop on Advanced Context Modelling, Reasoning and Management, UbiComp 2004.

Belk, R. W. (1988). Possessions and the Extended Self, In: Journal of Consumer Research, Vol. 15, No. 2, 139–168.

Bernhaupt, R., Mihalic, K., and Obrist, M. (2008). Methods for Usability Evaluation of Mobile Applications, In: Lumsden, J., (ed.), Handbook of Research on User Interface Design and Evaluation for Mobile Technology, Lumsden, J. et al. (eds.), Information Science Reference, Toronto, Canada, 742–755.

Blomberg, J., Giacomi, J., Mosher, A., Swenton-Wall, P. (1993). Ethnographic Field Methods and Their Relation to Design, In: Dchuler, D., Namioka, A., (eds.), Participatory Design: Principles and Practices, Erlbaum, New Jersey, USA.

Bowlby, J. (1969). Attachment, In: Attachment and Loss. Vol. I., London: Hogarth, London, UK.

Bradley, N., & Dunlop, M. (2005). Toward a multidisciplinary model of context to support context-aware computing. Human-Computer Interaction, 20 (4), 403–446.

Brandtzæg, Petter Bae & Asbjørn Følstad (2001). How to Understand Fun: Using Demands, decision latitude and social support to understand fun in Human Factors Design. In Helander, M., Khalid, H.M. and Tham, M.P. (eds.) Proceedings of The International Conference on Affective Human Factors Design (131–139). London; Asean Academic Press.

Brown, T., & Katz, B. (2011). Change by design. Journal of product innovation management, 28(3), 381–38.

Buchenau, M., J., F., Suri, (2000). Experience prototyping, In: DIS '00: Proceedings of the 3rd conference on Designing interactive systems, ACM, New York, USA.

Buxton, B. (2007). Sketching user experiences: getting the design right and the right design. Morgan Kaufmann.

Chow, R. (2006). Modelling and simulation of context-awareness, In: ANSS '06, Proceedings of the 39th annual Symposium on Simulation, page xi, IEEE Computer Society, Washington, DC, USA.

Crabtree, A. (1998). Ethnography in participatory design, In: Proceedings of Participatory Design Conference (PDC'98), CA: CPSR, Palo Alto, 93–105.

Csikszentmihalyi, M (1997). Finding Flow: The psychology of engagement with everyday life. New York: Basic Books.

Csikszentmihalyi, M. (1990). Flow: The Psychology of Optimal Experience. New York: Harper.

Davis, F. D. (1989). Perceived usefulness, perceived ease of use, and user acceptance of information technology. MIS Quarterly, 13, 319–339.

Dewey, John, (1980). Art as Experience. Perigee Books, New York, USA.

Dey, A., K. (2001). Understanding and Using Context, In: Personal and Ubiquitous Computing, Vol. 5, No. 1, February 2001, 4–7

Dey, A., K., Salber, D., Abowd, G., D. (2001). A conceptual framework and a toolkit for supporting the rapid prototyping of context-aware applications, In: Human Computer Interaction, 16, 97–166.

Dillon, A. (2001). User Acceptance of Information Technology. In Karwowski, W. (ed.). Encyclopedia of Human Factors and Ergonomics. London: Taylor and Francis.

Dix, A., Finlay, J., Abowd, G., D., Beale, R. (2004). Human-Computer Interaction (3rd Edition), Prentice Hall Inc.

Dourish, P. (2004). What we talk about when we talk about context, In: Personal and Ubiquitous Computing, Vol. 8, No. 1, February 2004, 19–30.

Dourish, P. (2006). Implications for Design, In: Proceedings CHI'06, Montréal, Canada, 541–550.

Dourish, P.; Belotti, V. (1992). Awareness and Coordination in Shared Workspaces. In: Proc. ACM Conf. on Computer Supported Cooperative Work CSCW '92 (Toronto, Canada, Nov.1992). New York, ACM Press, 107–114.

Engeström, Y., Miettinen, R., Punamaki, R.-L., (eds.) (1999). Perspectives on Activity Theory, Cambridge University Press, Cambridge, UK.

Fetterman, D., M. (1998). Ethnography: step by step, 2nd Edition, Thousand Oaks, Sage Publications.

Forlizzi, J., Battarbee, K. (2004). Understanding experience in interactive systems, In: Proceedings of the 2004 conference on Designing Interactive Systems, ACM, New York, USA, 261.

Gaver, B. Dunne, T.; Pacenti, E. (1999). Design: Cultural Probes, in 'Interactions' Ausgabe 1 (Jänner, Februar) / 1999; Association for Computing Machinery, New York, US, 21–29.

Hall, E. T. (1976). Beyond Culture. New York et al.: Doubleda.

Hassenzahl, M. (2003). The Thing and I: understanding the Relationship Between User and Product, In: Blythe, M., Overbeeke, C., Monk, A., and Wright, P., (eds.), Funology: From Usability to Enjoyment, Kluwer Academic Publishers, Dordrecht, 31–42.

Hassenzahl, M. (2004). The interplay of beauty, goodness and usability in interactive products, In: Human Computer Interaction, 19, 319–349.

Hassenzahl, M., & Tractinsky, N. (2006). User Experience – a research agenda. In: Behavior & Information Technology, 25(2), 91–97.

Heckhausen, H. (1980). Motivation und Handeln – Lehrbuch für Motivationspsychologie. Springer-Verlag, Berlin.

Herzberg, F., Mausner, B., and Snyderman, B.B. (1959). The Motivation to Work (2nd Ed.). New York: John Wiley & Sons.

Hinze, A. and Buchanan, G. (2005). Context-awareness in Mobile Tourist Information Systems: Challenges for User Interaction. International workshop on "Context in mobile HCI" in conjunction with MobileHCI 2005, September 19 2005, Salzburg, Austria.

Ho, J. and Intille, S. S. (2005). Using context-aware computing to reduce the perceived burden of interruptions from mobile devices. In Proceedings of the SIGCHI Conference on Human Factors in Computing Systems (Portland, Oregon, USA, April 02–07, 2005). CHI '05. ACM Press, New York, NY, 909–918.

Holtzblatt, K., Wendell, J., B., Wood, S. (2005). Rapid Contextual Design, A How-To Guide to Key Techniques for User-Centered Design, Morgen Kaufmann Publishers, San Francisco, CA, USA.

Hutchins, E. (1995). Cognition in the Wild. MIT Press, Cambridge, MA.

Jacko, J., A., Sears, A., (eds.) (2003). The Human-Computer Interaction Handbook, Mahwah, Lawrence Erlbaum, NJ, US.

Jordan, Patrick W. (2000). Designing Pleasurable Products: An Introduction to the New Human Factors, Taylor & Francis, London, UK.

Jung, Y., Persson, P., and Blom, J. (2005). DeDe: design and evaluation of a context-enhanced mobile messaging system. In Proceedings of the SIGCHI Conference on Human Factors in Computing Systems (Portland, Oregon, USA, April 02–07, 2005). CHI '05. ACM Press, New York, NY, 351–360.

Kaasinen, E. (2005). User acceptance of mobile services – value, ease of use, trust and ease of adoption. VTT Publications 566, dissertation, Tampere University of Technology.

Kaenampornpan, M., O'Neill, E. (2004). An integrated context model: Bringing Activity to Context, In: First International Workshop on Advanced Context Modelling, Reasoning And Management, UbiComp 2004.

Karapanos, E., Hassenzahl, M., Martens, J-B. (2008). User Experience Over Time, Work in Progress, In: Proceedings CHI 2008, ACM, New York, USA, pp. 3561–3566.

Koay, K., L., Dautenhahn, K., Woods, S., N., Walters, M., L. (2006). Empirical Results from Using a Comfort Level Device in Human-Robot Interaction Studies, In: HRI'06, (March 2–4, 2006, Salt Lake City, Utah), ACM, New York, USA, 194–201.

Larson, R., Csikszentmihalyi, M. (1983). The experience sampling method, In H. T. Reiss (ed.), Naturalistic approaches to studying social interaction, New directions for methodology of social and behavioral sciences, Jossey-Bass, San Francisco, 41–56.

Lieberman, H., Selker, T. (2000). Out of context: Computer systems that adapt to, and learn from, context, In: IBM Systems 39(3–4), 617–632.

Leavy, B. (2010). Design thinking-a new mental model of value innovation. Strategy & Leadership 38.3, 5–14.

Lowry, G. (2004). Translation and validation of the technology acceptance model and instrument for use in the arab world. In: ACIS 2004.

Massumi, B. (2002). Parables for the Virtual. Durham/London, Duke University Press.

McCarthy, J., Wright, P. (2004). Technology as Experience, The MIT Press.

Meschtscherjakov, A., Reitberger W., Lankes, M. Tscheligi, M. (2008). Enhanced Shopping: A Dynamic Map in a Retail Store, Ubicomp 2008.

Meschtscherjakov, D. Wiflinger, T. Scherndl, and M. Tscheligi (2009). Acceptance of future persuasive in-car interfaces towards a more economic driving behaviour. In AUI'09, 81–88. ACM, 2009.

Meschtscherjakov, A., Wilfinger, D., Tscheligi, M. (2014). Mobile Attachment – Causes and Consequences for Emotional Bonding with Mobile Phones. In Proceedings of the SIGCHI Conference on Human Factors in Computing Systems, CHI '14, ACM New York, NY, USA, 2014.

Mihalic, K. and Tscheligi, M. (2007). 'Divert: Mother-in-law' – Representing and Evaluating Social Context on Mobile Devices. Proceedings of the 9th international Conference on Human Computer interaction with Mobile Devices & Services (Singapore, September 9–12, 2007), MobileHCI '07.

Mihalic, K., Tscheligi, M., Huuskonen, P. (2006). Innovative Mobile Applications of Context, In: Proceedings of the 8th international Conference on Human Computer interaction with Mobile Devices & Services (Espoo, Finland, September 12–15, 2006), MobileHCI '06, ACM, New York, USA.

Muller, M., J. (2003). Participatory Design: The Third Space in HCI, In: Jacko, J., A., Sears, A., (eds.), The Human-Computer Interaction Handbook: Fundamentals, Evolving Technologies and Emerging Applications, Lawrence Erlbaum Associates Publishers, Mahwah/New Jersey, /London, UK, pp.: Lawrence Erlbaum Associates Publishers, 2003, 1051–1068.

Murer, M., Meschtscherjakov, A., Fuchsberger, V., Guiliani, M., Neureiter, K. Moser C., Aslan, I., Tscheligi, M. (2015). Contextual Interaction Design Research: Enabling HCI. In Human-Computer Interaction – INTERACT 2015–15th IFIP TC 13 International Conference, Bamberg, Germany, September 14–18, 2015, Proceedings, Part IV (2015), 621–623.

Nardi, B.A. (ed.) (1996). Context and Consciousness: Activity Theory and Human-Computer Interaction. Cambridge/MA, London: MIT Press.

Nielsen J. (1993). Usability Engineering, Morgan Kaufmann, San Francisco, USA.

Obrist, M., Beck, E., Kepplinger, S., Bernhaupt, R., Tscheligi, M., (2008). Local Communities: Back to Life (Live) through IPTV, Proceedings EuroiTV2008, Springer, Germany.

Osswald, S., Wurhofer W., Trösterer, S., Beck, E., Tscheligi, M. (2012). Proceedings of the 4th International Conference on Automotive User Interfaces and Interactive Vehicular Applications (AutomotiveUI '12), October 17–19, 2012, Portsmouth, NH, USA.

Pascoe, M. J. (1998). Adding Generic Contextual Capabilities to Wearable Computers. In: Proceedings of the 2nd IEEE international Symposium on Wearable Computers, ISWC. IEEE Computer Society, Washington, DC, 92.

Petrelli, D., Not, E., Zancanaro, M., Strapparava, C., Stock, O. (2001). Modelling and Adapting to Context, In: Personal and Ubiquitous Computing, Vol. 5/1, February 2001, 20–24.

Powdermaker, H. (1966). Stranger and Friend: The Way of an Anthropologist, W. W. Norton & Company Inc., New York, USA.

Raptis, D., Tselios, N., Avouris, N. (2005). Context-based design of mobile applications for museums: a survey of existing practices, In: Proceedings of the 7th international Conference on Human Computer interaction with Mobile Devices & Services (Salzburg, Austria, September 19–22, 2005), MobileHCI '05, Vol. 111. ACM Press, New York, USA, 153–160.

Rodden, T., Cheverst, K., Davies, K., and Dix, A. (1998). Exploiting context in HCI design for mobile systems. Workshop on Human Computer Interaction with

Mobile Devices. http://www.dcs.gla.ac.uk/~johnson/papers/mobile/HCIMD1.html.

Schmidt, A. (2001). Context-awareness, disappearing and distributed user interfaces, Experience, open issues and research questions, Presentation delivered at the Dagstuhl Seminar on Ubiquitous Computing.

Schneider, M., S., Kiesler (2005). Calling while driving: effects of providing remote traffic context, In: CHI '05, Proceedings of the SIGCHI conference on Human factors in computing systems, ACM. New York, USA.

Schrammel, J., Geven, A., Sefelin, R., Tscheligi, M., (2007). "Look!" – Using the Gaze Direction of Embodied Agents, In: CHI '07 extended abstracts on Human Factors in Computing Systems, April 2007.

Shedroff, N. (2001). Experience Design 1, New Riders, Indiana, USA.

Shilit, B., and Theimer, M. (1994). Disseminating active map information to mobile hosts. IEEE Network 8. 22–32.

Simmel, G. (1949). The sociology of sociability. In: American Journal of Sociology 55 (1949), 254–261.

Stone, D., Jarrett, C., Woodroffe, M., & Minocha, S. (2005). User Interface Design and Evaluation (The Morgan Kaufmann Series in Interactive Technologies).

Suchman, L., A. (1987). Plans and Situated Actions: The Problem of Human-Machine Communication, Cambridge University Press.

Tullis, T., Albert, B. (2008). Measuring the User Experience. Collecting, Analyzing, and Presenting Usability Metrics. Morgan Kaufmann.

Venkatesh, V., Davis, F. (2000). A theoretical extension of the technology acceptance model: Four longitudinal Field studies. Management science, 186–204.

Venkatesh, V., Morris, M., G., Davis, G., B., Davis, F., D., (2003). User Acceptance of Information Technology: Towards a Unified View, MIS Quarterly, 27(3), 425–478.

Venkatesh, V., Bala, H. (2008). Technology acceptance model 3 and a research agenda on interventions. Decision Sciences, 39(2), 273–315.

Wax, R., H. (1971). Doing Fieldwork: Warnings and Advice, University of Chicago Press, Chicago, USA.

Werner, O., G., Schoepfle (1987). Systematic Fieldwork, Sage, Newbury Park, CA, USA.

Westerink, J.H.D.M., Ouwerkerk, M., Overbeek, T.J.M., Pasveer, W.F., and de Ruyter, B., (2008). Probing Experience: From Assessment of User Emotions and Behaviour to Development of Products (Philips Research Book Series), Springer, Dordrecht, The Netherlands.

Wisneski, C., Ishii, H., Dahley, A., Gorbet, M., G., Brave, S., Ullmer, B., Yarin, P., (1998). Ambient Displays, In: Proc. CoBuild 1998, Springer, Berlin-Heidelberg, Germany.

Zimmerman, J., Forlizzi, J., Evenson, S. (2007). Research Through Design as a Method for Interaction Design Research in HCI, Proceedings CHI 2007, ACM, 493–502.

Ludwig Streit

Complextiy, acceleration, globalisation – A challenge for democratic governance

Abstract: *Democracy, as we know it, assumes transparency of the issues, sufficient time for their deliberation, and political sovereignty for implementation. Profound adaptations of democratic governance are called for to ensure legitimation and control under the auspices of complexity, acceleration, and globalisation.*

1 Introduction

It must have been around 1973 that Harald Zingl, my mentor in Graz, told me: "This kid Sepp Fröhlich is a very bright student; you will enjoy working with him."

And so it was. For a number of years we had a great time working in few body physics together, and Fröhlich's energetic and productive research made a successful career as a physics professor predictable.

But interesting biographies tend to be more complex. And so our scientific trajectories crossed once again in a different field, during the preparations of the Research Year "The Sciences of Complexity: From Mathematics to Technology to a Sustainable World" at the Center for Interdisciplinary Research ZiF of Bielefeld University. That research year was intended to cover the complexity impact in a wide range of fields. Fröhlich and his group proposed to cover, among others, information flows in regional innovation systems, models of environmental systems, and technology policy as an example of intervention in complex systems. Paradigmatic themes of research which are common today, but, remarkably, at that time, it was only due to Fröhlich's foresight (and charm) that they were finally admitted for inclusion in the research year.

With hindsight now it is patent that this research year would have been a much lesser success without his and his group's contributions.

For myself, preparations for this large-scale project in the final years of the past millennium made it clear that the emerging explosion of socio-economic complexity would lead to the existential challenge for democratic rule which we are now witnessing. The following considerations – essentially the transcript of a recent lecture at the Academy of Science in Lisbon (See also Streit 2015) – are intended to reflect this concern. They should not be read as a learned treatise from within Political Science, but as a stimulus to confront questions which have lain dormant in the public discourse for maybe too long, given the recent developments on both sides of the Atlantic.

2 The rise of democracy

Democracy – the rule by the people – was invented more than 2000 years ago to organise the social life of communities scattered around the Mediterranean, typically much smaller than present day Switzerland. Life then was dominated by the rhythm of celestial bodies: time was measured in terms of years, seasons, months – not much would happen in a day.

When industrialisation began to change the world and civil society was born, democracy was reborn with her, and was later contemplated by Winston Churchill[1], saying "Many forms of Government have been tried, and will be tried in this world of sin and woe. No one pretends that democracy is perfect or all-wise. Indeed, it has been said that democracy is the worst form of Government except for all those other forms that have been tried from time to time." Remarkably the first two sentences are quoted much less frequently than the last.

And in fact, any doubt about the superior quality of democratic government has been considered politically unacceptable.

How did the world look 200 years ago when democracy began to spread through Europe in the wake of the French revolution? Vast, endless, with open space to conquer and to explore or to get lost in; white, uncharted areas were abundant on the world maps. News of an Asian earthquake or tsunami would arrive at our shores many months after the event had happened and would be no more than an item of curiosity without much further effect on our lives.

Now, 200 years later, democratic rule has not only taken hold in most of the developed world, but has acquired the status of an ethical postulate. It is considered as an essential attribute of the superior social organisation and serves to "justify" military interventions such as e.g. those in the 2nd Iraq war.

And how does our world look now? Invented for those small city states of ancient Greece, democracy now has to face the challenges of the "global village" where news travel around the globe in fractions of a second, where events are tightly connected all over the world and effects can spread with lightning speed, and where, for example, the effect of a tsunami in Japan was able within days to impact the national German energy production portfolio in a profound way from which it has not recovered yet.

1 Speech in the House of Commons, 11 November 1947.

How did this happen? What innovation had the most important impact in shaping this world we live in today?

3 The rise of complexity

On August 12, 1981, IBM rolled out its first "Personal Computer"; digital computing entered the offices and living rooms. Mainframe computers had been around much earlier, they were born in the forties as a twin of the atom bomb. It is the global availability of personal computing power that made the difference. It made us all see and do things we would not have imagined earlier.

As an example, there is the famous Mandelbrot fractal. As we zoom into its finer and finer details, and so on ad infinitum, there appears a beautiful dream world of forms and colours. A complex structure indeed. My point here is: the mathematical formula behind it had been catching dust for a hundred years or so, but until the PC came, nobody had the slightest idea of the complex pattern that lay behind it[2].

It has been said that we lack an intuition for nonlinear phenomena and that perhaps *the emergence of such an intuition might mark a new intellectual awakening of mankind.*

This "new awakening" sounds like science fiction, something out of Kubrick's film "Space Odyssey 2001". Kubrick's film from 1968 was speculating that by 2001, a computer would attempt to get the upper hand over us. The takeover has not happened yet, but we do feel the threat. Even kids today have seen the Mandelbrot fractal somewhere or another, and terms from complex systems such as the famous "butterfly effect", have become buzzwords in the socio-economics discourse.

What is this impact of omnipresent computing power?

It marks a turning point of our quest for understanding in a wide range of pursuits – the end of reductionism. For two millennia, scientific problems had to be drastically simplified before being "understood". The dynamics of protons and neutrons is too hard to handle? OK, so let us describe them by a bunch of harmonic oscillators because those are simple enough to deal with. In physics, but as well in other disciplines such as e.g. economics, simplifications and massive complexity reductions were all over the place. Collective behaviour was often beyond the reach of reliable modelling.

---

2 Hear Benoît Mandelbrot himself about the wonders of this computer-based breakthrough: http://www.ted.com/talks/benoit_mandelbrot_fractals_the_art_of_roughness/transcript?language=en#t-586141.

Now, with our exponentially growing computing power, complex systems, and in particular the collective behaviour of large systems can be studied, and this New Awakening, an intuition for the strange behaviour of complex systems, is rapidly developing. The tools were there at the dawn of the new millennium, the time was right.

But the computer not only changed our understanding of the world, it has changed our own world, dramatically.

With millions of PCs linked in the Web, our cultural, economic and social world has become a closely connected global network. Information flies at almost the speed of light. The effect of local disasters or interventions is felt almost instantaneously around the world.

All of this has become an incredible complex socio-economic system, with enormous risks of spontaneous destabilisation.

3.1 So what is complexity?

As an example, let us have a look at the Malthusian growth of populations. In this model, the new population is assumed to be proportional to the size of the previous generation, where f is the fertility rate and whenever the fertility f is larger than one, the population will grow; in fact, it then grows exponentially.

$$P_{n+1} = f \cdot P_n$$

But the real world is more complex: the effect of competition is felt when populations become big and can be described by

$$P_{n+1} = f \cdot P_n - c \cdot P_n^2$$

The new population is proportional to the size of the previous one, minus the effect of competition for limited resources, encoded in the parameter c.

Now we expect that the population will saturate at a limiting size, before the quadratic second term becomes too big. But this is not all that can happen. There can also be oscillations, and even chaos as time progresses from one generation to the next.

We should note here two important features of this by now classical model:

1. All these scenarios can occur if one varies the fertility rate f or competition c just very slightly! In this sense the evolution of populations becomes *practically unpredictable* since we can never be sure of those parameters with the necessary precision to exclude one or the other scenario.
2. Even if the formula itself is extremely simple, its less obvious consequences were only readily available in the age of the PC.

3.2 What then is complexity?

A concept of such generality necessarily has several definitions. They will be context dependent, but not only. In the context of computation for example, complexity may mean a very complex computer program, or one that takes a long time to run, or one that uses a lot of memory, or one that is simple in all these respects but produces surprising results, like our population model.

There are many characteristic features which are of relevance in our present context. One might for example mention the intermittency phenomenon where complex systems, after a prolonged period of regular evolution, suddenly show a different, even chaotic behaviour, without any external intervention. Or one might draw attention to the rich and often counter-intuitive features of complex networks and their relevance for modern urban planning, among others.

But to keep things simple here we will just focus on two general properties of complex systems:

- Their evolution is difficult to predict because they depend heavily on fine details of information that may be beyond our control. Weather and climate are well known examples of this difficulty (this is the famous *Butterfly Effect*).
- A variant of this is the difficulty to predict the reaction of complex systems to even small external interventions.

4 Consequences

From Linear Planning to Adaptive Management. Crucially important is the lesson that – all the way from the micro-economic, business level to the macroeconomic level of national and international policy making – planning and strategies have to be continuously readjusted to counteract the surprises of the butterfly effect and of random exterior influences.

According to Lenin, the whole national economy should be organised on the lines of the postal service. In practical Soviet terms, this meant governance through centralised plans, based on an equilibrium model of the economy. The result is well known. When East Germany was irreversibly bankrupt, the Berlin Wall fell and the communist bloc crumbled.

So are we "Lost in Complexity"? J. Gleick writes: "… a new generation of scientists has come along… They know that a complex, dynamical system can get freaky. They know, when it does that, that you can still look it in the eye and take its measure." (Gleick 2008)

In a modern highly complex economy, long term planning must clearly be replaced by continuous monitoring and feedback loops. The – admittedly rather sophisticated – techniques to compensate random disturbances on the fly go back to Cold War technologies, they were invented under the name of optimal control or dynamic programming, mainly to keep ICBMs on target during their flight so that they would surely hit New York or Moscow in spite of any in-flight perturbations. Economists and more generally people from the humanities tend to be allergic to the notion of control theory – it sounds too repressive to their tender ears. But now economists have discovered a nice new label, that of "adaptive management", it is the same thing, and now increasingly popular, a recent buzzword in economics.

4.1 Acceleration and globalisation

Of course the problems of socio-economic complexity and the need for their adaptive management have been extremely aggravated by globalisation and acceleration. While economic news used to need months to travel around the globe before the advent of steamships, information now flies around the globe with the velocity of light, and hectic reactions are a potential source of serious instability

As an example, financial transaction strategies with the potential to produce crises of a global dimension take place in milliseconds, powered by computer algorithms, without the possibility of adequate human intervention. There have been instances where a large scale financial crash was only avoided by literally pulling the plug of those computers at the New York stock exchange. New York banks place those computers right next to the Chicago stock exchange to shorten the transmission time by some tiny fractions of a second, gaining a multi-million dollar advantage in this way.

As one of the consequences of globalisation, Asian workers are now in a very direct competition with their colleagues in the Americas or Europe, the work force has been globalised, and employers go for the cheapest offer worldwide. In the Manila metropolitan region alone, there are more than one million employees working in call-centres for American, Australian and other companies and even universities.

4.2 Clouds on the horizon

Like with other revolutionary inventions, there is a definite dark side to our exponentially increasing computing and data-handling power.

Paradoxically, the net and digital data-handling have produced a new kind of reductionism: to see this, compare the rich and manifold traditional meaning of the word friendship with its reduction to a mouse click in the Facebook world, or the rich and unique spectrum of a Stradivari violin with its reduction under MP3[3].

Then there is the loss of privacy and freedom of choice: You pay for the web with the loss of your privacy, and the web then defines your identity. Google et al. will decide what you will want to see, what to know, what to buy.

For those who still adhere to the belief or at least the hope that our new world will be bathed in the bright light of global transparency, unlimited access to knowledge for everyone on the globe, here is an exercise to test this hypothesis on unlimited access:

These days our profile, our digital avatar in the web, is getting more and more complete at a rapid rate.

On the web, there are our Credit Rating, Health Status, Personal Reliability, Sexual Preferences, People We Dislike, Gadgets We Would Want To Own, Some Stupid Thing We Did 10 Years Ago, Political Leanings and Beliefs, Our Body Odor (if any), Professional Competence and Performance,… to name just a small random sample.

Here is now the exercise to test the true quality of that new transparency paradigm: Try to

1. Obtain your stored profile from the Lords of Big Data.
2. If necessary, correct or erase them.

So much for the Brave New World of total transparency: the individual has lost his privacy and secrets, while his digital counterpart is a secret, closely guarded for unknown purposes.

And we have to be concerned with a cyberspace beyond the reach of the law. Google and Microsoft were reported to be planning to move activities to the "off-shore". Larry Page, co-founder of Google, said in 2013:

> *„There's many, many exciting and important things you could do that you just can't do because they're illegal… we should have some safe places where we can try out some new things and figure out what is the effect on society, what's the effect on people…"* (Döpfner 2014)

Are we being manipulated? A year ago or so, it would have been considered Science Fiction that Social Media manipulations would determine the outcome of elections. Contrary to Hillary, Trump did not commit himself to identifying and addressing the main national issues. He did not need to: instead, his campaign was

3 A dramatic account by one who is considered "the father of virtual reality technology" can be found in **J. Lanier (2010)**.

based on an exploration of voter profiles stored in Big Data. Then via the Social Media, he fed individual groups with what they wanted to hear. This way he ran a cheap and, as we know, surprisingly effective campaign.

Only a few years ago, governments were at a loss to control the implosion of the financial markets, these days the German government has expressed its concern that this year's federal elections in Germany will be decided by a group of hackers in Moscow or St. Petersburg.

It is clear that only a massive government intervention can preserve our privacy, and – much more dramatic – our freedom as individuals, freedom of choice, but also the right to a self-determined individuality. What is the use of your going to a job interview if the personnel managers think they know everything from the internet? And the cloud does not forget, and will not be corrected where it is wrong.

Unfortunately, it is also abundantly clear that national governments are light-years away from the necessary competence to deal with the Lords of the Data on an equal footing.

Acquisition of this competence, and the will to employ it for effective regulation before the first Malevolent Lord takes over, is a race with uncertain outcome. If humanity loses this race, its definition will have to be rewritten, as an army of normalised ants following the orders of an unaccountable elite in the clouds. North Korea might end up looking idyllic in comparison.

Like nuclear energy half a century ago, Big Data is an innovation of high potential but which will be deadly if not harnessed. Massive mathematical, technical, but also social, legal research outside the confines of Silicon Valley, and then a sharpening of public awareness are of the essence.

5 Democracy – Rule by the People?

The democratic decision-making process requires:

- Transparency of the issues
- Transparency of, and confidence in the effectiveness of the strategies available
- Enough time to present and discuss these issues and strategies, and often to test their legal issues in court
- On the other hand, nowadays, an adaptive political management is needed, responding quickly and promptly to unexpected threats and challenges

We lack sufficient regulatory capacity to deal with banks "too big to fail", to deal with an economy 4.0 that does not know borders and evades national laws.

We have to cope with the anti-intuitive behaviour of complex systems: they are highly non-linear – double in will not produce double out. Raising a tax may produce less instead of more tax revenue because business moves to other markets. Modifications in the health care system may produce utterly unforeseeable effects, and so on and so forth everywhere in government. Politicians will cite their favourite experts, but even experts are at a loss for reliable predictions. To handle the financial crisis competently the German government had to resort to the advice of Germany's leading banker Joseph Ackermann. This was admittedly a blatant case of conflict of interest, but the government simply did not have the necessary competence to understand the problems and necessities and to arrive at an independent judgment speedily. In 2009, time of the global banking crisis, the German economics minister asked a cabinet of specialised lawyers to draft a law concerning the administration of ailing banks. Criticised for giving this task to a law firm with important ties to international banking and the ensuing risk of conflict of interest, he protested that the ministry itself simply did not have the expertise to handle this highly complex matter.

In view of present-day complexity and acceleration, it is no surprise that the lifespan of regulations becomes shorter and shorter before they need to be rewritten[4].

Some of the Challenges: The issues of demography, public debt, sustainable energy, ecology, protection of privacy, terrorism, massive immigration are undoubtedly giant questions, touching upon the determinants of society, – and even the man in the street begins to sense the incapacities of the democratic political system.

Social disparity in particular is a sleeping monster, the growing gap between the few rich and the many poor, power and wealth concentrated in the grip of a few, and the masses disenfranchised – this is clearly an unstable trajectory and clearly jeopardizes the democratic process more and more.

After the collapse of the Lehman Brothers investment bank in New York, an ultra-fast and high intensity dialogue between governments and central banks was needed to prevent a global collapse of the financial system.

Then there was the crisis of public debt: bankruptcy of Greece and other countries was avoided through loans and guarantees from richer EU countries. These allocations amounted to a significant fraction of their national budgets.

Most recently, Europe was literally overwhelmed with a seemingly uncontrollable wave of immigration.

---

4 For more on this, see e.g. J.-C. Michéa (2009).

In all these cases parliaments were – at best – informed by the executive that the situation was far from transparent, but large financial commitments were without alternative, and were so urgent there was no time for any proper discussion on them. Very complex and dramatic issues, and parliaments are told that there is no time to discuss them at any length.

It should be recalled that the authority over the national budget, about how to allocate taxpayer money, is the central and most important privilege of Parliaments. Or was it…?

5.1 Post-Democracy

No wonder that the public feels at a loss with regard to the big political questions. Low voter turnout is the consequence.

Colin Crouch observes "*… Even if elections take place and continue to influence governments, the electoral debate is a tightly controlled show, rival groups led by experienced professionals in the techniques of persuasion practice on a limited number of questions selected from these groups. The mass of citizens plays a passive, acquiescent, even apathetic role, merely reacting to the signals it receives. Apart from the spectacle of the election campaign, policy is decided in private by the integration between elected governments and elites…*". (Crouch 2004)

One just has to look at the propaganda before elections. The enormous gap between the complexity of the current problems, and the extreme simplicity of the "slogans" of political parties highlights the helplessness of the political class to communicate properly with the electorate, and effectively dissuades the people from exercising its democratic right of voting. (Habermas 2015)

Public reactions to this disillusionment of the electorate for too long have tended to be politically correct: characteristically, the ensuing discussion focused almost entirely on somehow keeping democracy going by working on the symptoms. The Secretary General of Germany's oldest and most venerable political party recently proposed to locate polling stations in shopping malls or big railway stations and to keep them open for a week to revive voting interest, instead of facing the delicate question whether the democratic process in its present form might be fundamentally inadequate.

But, to quote Bob Dylan (Nobel Prize 2016) "*… the times they are a-changin'…* "

Only a few years ago, suggestions of Russian intervention in an American election would have been shrugged off as science fiction. How will we cope with campaigns that, unseen to the general public, are discretely wooing different interest groups with different messages and promises?

These days the German authorities declare their concern with the threat that the upcoming federal elections might be decided by hackers at the behest of some foreign government.

A few days before (!) the recent US elections, the New York Herald Tribune ran a one page analysis of the Trump and Clinton campaigns, expressing the worry that contrary to Trump, Clinton might have made a strategic mistake since she had been highlighting political issues and was elaborating concrete measures to cope with them, instead of just celebrating herself in the rallies.

The threats to democracy are rapidly becoming so much more patent that unheard of proposals for its rescue are now entering the public discourse, such as e.g. David van Reybrouck's "Against Elections", an argument for sortation as a *defence* of democracy (Reybrouck 2016). I am not suggesting that van Reybrouck's proposal of sortation is a cure-all, my point is rather to underscore that a well-informed debate on the challenges of democracy is entering the public discourse, instead of leaving the arena to the populist movements which are springing up all around as a brainless reaction to the complexity of our world.

5.2 Democracy – Two imperfect alternatives

There is Singapore with no natural resources, and a delicate balance between three ethnic groups of very different cultures – Chinese, Malay, and Indian. Rising from great poverty and massive unemployment, it is now one of the richest societies in the world, with very low crime rate and corruption, reasonable human rights, but with a very limited political participation.

Then there is China, often cited for human rights deficits, and with extremely low political participation, but quickly becoming an industrial superpower, global, financial and military, feared by its neighbours of becoming the bully in the region, but also with an impressive record in eliminating poverty. An 88 % drop of the number of people living in poverty within the 23 years from 1978 to 2001 (Jacques 2009) shows how China is winning the war against poverty, and thus the approval and gratitude of many.

These two models cannot and should not be copied. But it would also be a big mistake not to study them carefully with regard to their important technocratic successes.

In any case, these two examples highlight the fundamental conflict between efficiency and legitimation.

But they also point to the necessity for government, and politics more generally, to learn. Leading members of the Singapore government have a remarkable competence regarding the problems and challenges inherent in the complexity of their society and make strong efforts to integrate those insights in their strategies.

On the legislative side, and in a more democratic setting, the Austrian parliament is setting an example: they had invited the Innovation Systems unit of the Austrian Institute of Technology to coach them on complexity issues and have recently asked that team for a second round of these debates…

6 Summarising: Questions, doubts and challenges

The most important feature of democracy, the one that is essential to be maintained and protected, especially in these troubled times, is of course its control of the political process. The problem we need to address here when we look at our role as voters is threefold:

- Ex ante control is insufficient in view of *Complexity*, which entails intransparence of the issues and unpredictability of outcomes.
- Ex post control often comes much too late in view of *Acceleration.*
- Both have limited impact because of the transnational nature of many issues due to *Globalisation*.

Can the democratic process effectively deal with increasingly important issues which are – at best – understood only by a small group of experts, taking into account that they occur so fast that there is no time to consult the legislative, much less the electorate, and that they require a considerable amount of continuous adaptive management and imprevisible, ad hoc government intervention?

Or else:

Is it advisable – or is it possibly too risky – to seek and study possible adaptations or alternatives of the democratic process as we know it?

What could these modifications be? How can they ensure the quality of the governmental system and civic control of its performance? *What do you think?*

Acknowledgements: My thanks go to my late colleague Glória Cravo, Funchal, for her encouragement, and to Josef Fröhlich, Vienna, for decades of friendship and enlightenment.

References

Crouch, C. (2004). Post-Democracy. Polity Press, Cambridge.

Döpfner M. (2014). An open letter to Eric Schmidt from Mathias Döpfner, http://www.axelspringer.de/dl/433625/LetterMathiasDoepfnerEricSchmidt.pdf, accessed March 7, 2017.

Gleick, J. (2008). "Chaos: Making a New Science". Revised edition, Penguin.

Habermas, J. (2015). Critique et communication: les tâches de la philosophie. Esprit 2015/8 40–54.

Jacques, M. (2009). When China Rules the World. Penguin, NY, 162.

Lanier, J. (2010). You Are Not a Gadget. Knopf, NY.

Michéa, J.-C. (2009). The Realm of Lesser Evil. Polity Press, Cambridge, UK.

van Reybrouck, D. (2016). Against Elections – The Case for Democracy. Random House.

Streit, L. (2015). Complexity, Acceleration, Globalization – A Challenge for Democracy. Interdisciplinary Studies of Complex Systems 7, 90–98.

Authors

Laurent Bergé is a post-doc researcher at the Centre for Research in Economics and Management at University of Luxembourg. He received his doctoral degree in Economics from University of Bordeaux. In his research he focuses on network dynamics, determinants of innovation and the economics of science.

Eva Buchinger is a scientist at the Austrian Institute of Technology, specializing in innovation and systems theory. She is, among other roles, board member of the ISA Research Committee on Sociocybernetics, lecturer at the Technical University of Vienna and consultant for the Austrian Government and EU Institutions.

Wolfgang Czerny graduated in Physics and Mathematics and was active in nuclear physics research. He studied systems theory and cybernetics and until his retirement worked as a scientist at the AIT Austrian Institute of Technology with a focus on innovation research and complex systems.

Bernhard Dachs is Senior Scientist at the Center for Innovation Systems & Policy of AIT. His areas of expertise are the economics of innovation and technological change, in particular the internationalisation of R&D, innovation in services, and the analysis of technology policy.

Michael Dinges is an economist and thematic coordinator of the research area Innovation Governance at AIT Austrian Institute of Technology. He has long lasting experience in designing and managing evaluations at European and national level with a particular focus on research and innovation policies and science-industry collaborations.

Martina Dünser is a researcher at the AIT Center for Innovation Systems & Policy. Her research focuses on modelling and analyzing innovation processes, especially the creation of knowledge in regional innovation systems by means of agent-based simulation.

Margit Fischer is the initiator and President of the Association ScienceCenter-Network. Her professional background is in textile design, weaving and history of art.

Susanne Giesecke is a senior scientist and thematic coordinator of "Foresight" at the Austrian Institute of Technology's Center for Innovation Systems & Policy. A political scientist by training, she presently works on qualitative innovation research and Grand Challenges, technology assessment; evaluation of R&D programmes as well as foresight.

Barbara Heller-Schuh works as a scientist at the Center for Innovation Systems & Policy of AIT Austrian Institute of Technology since 2002. She graduated in History, German Language and Literature studies and in Knowledge Management. Her main research interests cover research and innovation policy, innovation networks and higher education research.

Karl-Heinz Leitner is Senior Scientist at the Center for Innovation Systems & Policy at the Austrian Institute of Technology and Professor for Innovation Management at the University of Graz. His main research interests cover new forms of innovation, foresight, innovation policy and the valuation of intellectual capital.

Simon Martin is enrolled as a PhD candidate in the Vienna Graduate School of Economics (VGSE) at the University of Vienna. He holds a Master in Economics and a BSc in Business, Economics and Social Sciences with Management Information Systems major. His research interests are microeconomics, game theory and industrial organisation.

Manfred Paier is a scientist at the AIT Center for Innovation Systems & Policy. He holds a diploma in Theoretical Physics from the University of Graz, and performed post-graduate studies in energy as well as industrial economics. His research focuses on knowledge production, research networks and innovation dynamics covering a variety of industrial sectors.

Rupert Pichler is Head of the research and technology funding department at the Austrian Ministry for Transport, Innovation, and Technology. He holds a master's and a doctoral degree in history and was a researcher at the Austrian Academy of Sciences. He is the author of books, book chapters and journal articles on research policy and economic history.

Wolf D. Prix is the Design Principal and CEO of COOP HIMMELB(L)AU. Amongst others, he taught at the Southern California Institute of Architecture (SCI-Arc), Harvard Graduate School of Design, Columbia University, UCLA University of California and Yale School of Architecture. From 2003–2012 he was a Vice-Rector and Head of the Institute of Architecture at the University of Applied Arts in Vienna.

Since March 2003, **Ingolf Schädler** has been serving as Deputy Director General for Innovation at the Austrian Ministry for Transport, Innovation and Technology (BMVIT). In this function he is responsible for the development and implementation of the Ministry's technology policy, its strategic technology programmes and the related R & D infrastructure.

Petra Schaper-Rinkel has been a Senior Scientist at the AIT Austrian Institute of Technology since 2009, where she drives principally EU projects on science, technology and innovation policy and foresight.

Doris Schartinger is a scientist at the AIT Austrian Institute of Technology in Vienna. Her primary research focus is on the economics of innovation and technological change. She has been involved in a number of research projects ranging from theoretical and empirical studies of innovation, to managerial and policy dimensions of these issues.

Thomas Scherngell is a senior scientist at the AIT Center for Innovation Systems & Policy, and lecturer at Vienna University of Economics and Business (WU). He holds a habilitation in Economic Geography at WU, and a doctoral degree at University of Vienna. His research focuses on economics of innovation and innovation networks.

Edgar Schiebel studied mechanical engineering at the Technical University of Karlsruhe, holds a PhD from the technical university of Vienna and is a senior scientist at the Austrian Institute of Technology. His academic activities cover lectures on Technology Foresight, Business Intelligence and Science Mapping at the universities of Ulm and Graz and the supervision of scientific theses and publications in several journals.

Barbara Streicher works as the Executive Manager of the Association ScienceCenter-Network since its foundation in 2005. She graduated in Molecular Genetics and has a long-term experience in science communication and research. Her work focuses on hands-on science center activities for various publics.

Ludwig Streit is professor emeritus at Bielefeld and Funchal, honorary professor in Graz and IIT, Iligan, member of the Academy of Science, Lisbon. His research topics include stochastic analysis and quantum physics. He has managed and counsels research institutes in Europe and Asia.

Manfred Tscheligi is Professor of Human-Computer Interaction at the Center for Human-Computer Interaction, University of Salzburg; and Head of Center for Technology Experience at AIT in Vienna. His main research focus is on the synergy of actual and upcoming digital contexts and the very essence of technology experience and human-computer interaction modalities in these contexts.

Iris Wanzenböck is assistant professor at the Copernicus Institute of Sustainable Development at University of Utrecht. She received her doctoral degree from Vienna University of Economics and Business (WU), in cooperation with the AIT Center for Innovation Systems & Policy. Her main focus of research is on the impacts and determinants of innovations networks.

Matthias Weber is Head of Center for Innovation Systems & Policy at AIT Austrian Institute of Technology in Vienna. He has a background in process engineering, political sciences and economics, and worked for more than twenty years on innovation systems, strategic foresight, and STI policy.

Georg Zahradnik is a Scientist at the Austrian Institutes of Technology, Center for Innovation Systems & Policy. Over the past five years, his research focus was on the internationalisation of research activities, public research funding and sectoral innovation systems.